DESIGN FORMULA
APPRECIATION OF SHOW HOUSES
设计方程式——样板空间风格赏析

深圳视界文化传播有限公司　编

中国林业出版社
China Forestry Publishing House

PREFACE
序言

CREATE YOUR OWN SPACE
打造专属于你的空间

I enjoy life in my own way, and I never compromise to do ordinary design.

For me, design has always been a magical thing. My design inspiration comes from the beautiful nature which gives us more fun. Taking advantage of the situation, I pay attention to the harmonious relationship among the nature, people and buildings, and then have a flexible and adaptable creation. There is nothing like nature which creates such a soft and vivid color. It is she who taught us to use subtle differences to make the original limited scope release the greatest splendor in the integration of space decoration.

Views on the personality and aesthetics vary from person to person, so the designer needs to observe the details of customer's life to make each non-professional house-owners can clearly understand the importance of design and realize that they can offer what kind of strength and luster for their home. We do not blindly provide what customers want, but try our best to offer the most professional, suitable, appropriate things that they perhaps never dream of. When the customer gets it, he/she will realize that this is what he/she has always wanted. This is the direction what we have been pursuing and insisting in design.

Through years of hard decoration design experiences, I deeply understand that the design will lack of the inner feelings and cannot fully express the unique way of life space if only adjust the structure. Therefore, in order to design perfectly and follow my heart, I founded my own soft decoration design studio -- Hangzhou JULIE Soft Decoration Design in 2013, began to do the full case design and overall planning of well-decorated houses, which seamlessly combined hard and soft decoration. Through the control of space structure, material selection and color collision, I explore charming aesthetics that suits each independent space, and inject different fresh breath of life into each new home as much as possible!

Sometimes people think that my use of color is too bold, but I have always insisted that we should not be restricted in the charm of the collision of any color itself, so that I try to create the infinite sparks in a limited space for the integration of aesthetics and practicability. Maybe I am a designer who prefers to use color to highlight myself. I think everyone has their own unique temperament, and what we have to do is to create their own personalized space.

  用自己的方式去热爱生活，坚持做不将就的设计。

  对我来说，设计一直是一件神奇的事情，它的灵感来自于美丽的大自然，赋予我们更多乐趣。因势利导，讲究自然、人、建筑之间的和谐舒适关系，再进行富有灵活性和适应性的创造。没有什么能像大自然一样创造出如此柔和饱满的色彩，她教会我们利用细微的差别来整合装扮空间，将原本了无生趣的局限范围释放出最大的光彩。

  由于每个人的个性与审美观点并非一样，需要设计师通过对客户生活细节的观察，让每个非专业的业主通过我们，可以清晰地懂得设计的重要性，清晰地看到自己可以赋予这个家什么样的力量和光彩。我们并不是一味地给予客户他想要的，而是及时并竭尽所能地给予客户最专业、最适合、最贴切，同时也许从未梦想过的东西。而当客户得到它之后，他会认识到原来这才是他一直想要的东西。这便是我们设计一直在追求和努力坚持的方向。

  多年的硬装设计工作让我深刻体会到，设计仅仅通过对结构的调整是远远不够的，它会缺少内在的情感，而且不能完整地表达出空间独特的生活方式。于是，为了将设计更完整地落地，听从自己内心的声音，我于2013年创立了自己的软装设计工作室——杭州JULIE软装设计，开始做全案设计及精装房的整体软装配置，将硬装与软装无缝结合。通过对空间结构、材质选择与色彩碰撞的把控，探究出适合每个独立空间的魅力美学，尽可能为每个新家注入不一样的新鲜的生活气息！

  有时候，人们会认为我对色彩的运用太过大胆，而我一直坚持我们不该禁锢任何色彩本身的碰撞魅力，让它们尽量在有限的空间里创造出无限的火花，将美观与实用融为一体。我可能是一个比较喜欢用色彩来彰显自我的设计师，我认为每个人都有属于自己独特的气质，我们要做的就是打造专属于他们的个性空间。

HANGZHOU JULIE SOFT DECORATION DESIGN / XIONG ZHENZHEN
杭州JULIE软装设计 / 熊真真

CONTENTS
目录

BRITISH AND FRENCH STYLE | 英法风格

008	*Golden Period*	鎏金岁月
020	*A Romantic French Fantasy Journey*	浪漫法式的奇幻之旅
030	*Dwellers inside the Landscape*	山水深处有人家
040	*Elegant Residence, Art Garden*	优雅之居 艺术花园
048	*Luxurious Residence, Elegant Life*	奢华居室 优雅生活
058	*Luxurious and Elegant Fashion Style*	华贵典雅 时尚气质

MODERN STYLE | 现代风格

070	*Elegance and Luxury Can Have its Own Unique Label*	雅奢——可以有自我独特的标签
076	*Toughness Lies in Entirety, Softness in Details*	大处见刚 细部现柔
088	*Splendid Color*	流光溢彩
096	*In Search of Lost Time, Low-key Luxury*	追忆似水 奢华低调
104	*The Integration of Elegance, Luxury and Comfort; Function and Beauty*	优雅奢适 功能与美的融合
110	*The Storyteller*	讲故事的人
116	*Modern Interpretation of Fashion Orient*	时尚东方的现代演绎
122	*Forest Holiday: Great Beauty Lies in Simplicity*	森林度假风：简约之中存有大美

NEO-CHINESE STYLE | 新中式风格

132	*Mountain House*	居山院
138	*Melodious Ancient Rhyme*	悠悠古韵

A Zen Style Retreat	世外禅居	**146**
The Heart of Grass and Wood, the Home to Zen Flavor	草木为心 禅意之家	**156**
Leisurely Landscape	悠然山水间	**162**
Oriental Impression	印象东方	**170**
Home of Art Collection	艺术收藏之家	**178**

NEO-CLASSICAL STYLE | 新古典风格

Low-key Interpretation of Luxury Temperament	低调演绎奢华气质	**192**
Extreme Luxury, the Taste of Home	极致奢雅 演绎家的味道	**208**
Fairyland in the World, Flower and Bird Paradise	人间仙境 花鸟天堂	**220**
Great Harmony	大同	**226**

AMERICAN STYLE | 美式风格

Beating American Feelings	跳动的美式情怀	**242**
An American Mansion of Free Design	追求自在设计的美式美宅	**258**
Notting Hill	诺丁山	**264**
A Gentle Remembrance	一段温柔的忆想	**274**
Fashionable Pulsations in Shanghai, Listening to Free Voices	时尚魔都脉动 聆听自由心声	**282**

SOUTHEAST ASIAN STYLE | 东南亚风格

Dazzling Rainbow	彩虹之炫	**292**
Home of Rainforest	热带雨林之家	**302**
Simplicity out of Complexity, Deep Love for Nature	出繁入简 厚爱自然	**312**

BRITISH & FRENCH STYLE
英法风格

英法风格 BRITISH & FRENCH STYLE

CHONGQING YOKY DESIGN
重庆于计设计

项目名称：江润紫云岭别墅样板间

项目地点：重庆

项目面积：300m²

主要材料：大理石、金属、软包、挂画等

摄 影 师：刘星昊

Golden Period
鎏金岁月

DESIGN CONCEPT | 设计理念

This case in French country style uses warm and simple colors, plain furniture, approachable design factors, casual and natural decorations to create a refreshing sensory effect. With no exaggerated form nor strong visual impact, the case uses such gentle colors as white, azure, pale gold combined with elegant and delicate home furnishings and decorations, making the entire space present a tranquil and romantic scene. The carpet with stretching flowers and leaf branches in the living room seems to exude flower fragrance. In terms of accessories, the basic color is white and blue interspersed with gold. With the French style crystal chandeliers, the space looks bright, transparent, majestic and noble, showing the host's romance and elegance.

法式乡村风格完全使用温馨简单的颜色及朴素的家具，使用令人备感亲切的设计因素，随意、自然、不造作的装饰及摆设方式，创造出如沐春风般的感官效果。没有夸张的造型，亦无强烈的视觉冲击，纯白、天青、淡金……这些温柔的颜色静静融合，结合线条优雅、造型精致的家居和装饰品，整个空间呈现出一幅恬淡、浪漫的画卷。客厅的地面上铺着舒展着花叶枝条的地毯，花香仿佛已在鼻尖。配饰上，以白色、蓝色为主色调，金色为装饰，整体空间再搭配上法式风格的水晶吊灯，看起来明亮、通透、非凡气度、贵气逼人，而且无一不展示出主人的浪漫、优雅。

Formula ｜ 方程式 ①

Porcelains are the most dazzling decorations in this house in which blue-and-white patterns are applied to the vase, tea sets, wall decorations and other elements to beautify the space. Meanwhile the cloth, floor, furniture and others share the same color or pattern of the porcelain, which complement each other.

　　瓷器，是此居室最亮眼的装饰品。花瓶、茶具、墙上的装饰品等，都用了青花纹饰美化空间。布艺、地板、家具等都是用的青花瓷的色彩或者是花纹，它们相互交融、相得益彰。

Formula ｜ 方程式 ②

The golden copper rim decorations meet the eye everywhere in the house, for example, the furniture, wall and ground edge are embellished with exquisite and elegant copper frames, which are low-key and elegant, noble yet not showy, enriching the color of the space and making the space more delicate.

　　金色铜条边框的装饰在居室中比比皆是。在家具、墙面包括地面边带的细节上，饰以精细优美的铜条边框做装饰，低调中彰显高雅，豪华不俗、独特不喧。它们不仅丰富空间的色彩，而且使空间更加细腻，并且永不落时。

英法风格 BRITISH & FRENCH STYLE

XIONG ZHENZHEN
熊真真

项目名称：昆仑公馆

设计公司：杭州JULIE软装设计

项目地点：浙江杭州

项目面积：400m²

主要材料：大理石、壁纸、布艺、水晶等

A Romantic French Fantasy Journey
浪漫法式的奇幻之旅

DESIGN CONCEPT | 设计理念

This case is named after the royal family with a noble temperament permeating in the air. Standing in the extravagant air of the greenhouse palace, she is noble and pure with her romance and love deeply carved into the bones and soul. Unlike the classical French style, the luxurious and stylish furniture decorations infuse gorgeousness and delicacy into the space, presenting the details different from the common French style. It is the designer's persistence to inherit the French feelings.

她的故事以皇室命名，高贵的气质在空中弥漫，她的身影立在温室殿堂的奢靡空气里，高尚纯净。她的浪漫深入骨髓，她的爱刻入灵魂。不同于经典的法式风情，奢华时尚的家具摆件，将华美与精致融入空间。娓娓讲述关于法式和一贯认知里不一样的细节，那些法式的情怀，我们执守。

The two spaces with exquisite furniture style and rich colors are linked together by a fly-wing-to-wing ornament, full of vigor and vitality.

相互联系的两个空间，精致细腻的家具款式，通过比翼双飞的装饰摆件将两者联系贯穿，饱满的色泽，富含生机和活力。

With the original exquisite stone as its basic material, the case uses dreamy blue as the primary color of decoration, which is applied to curtains. Together with the highlight color to restore the modern French aesthetic and the blue gray color of the fireplace, it reduces the sense of distance and injects an ever-fount gentle temperament to the space. The bronze decorative mirror not only enhances the level feeling, but brings a comfortable visual effect. Complemented with the copper crystal lamp, it shows a noble temperament that can not be ignored in the space.

在原有精装石材的基础上，加以梦幻蓝为装饰主打色，无论是应用于窗帘，还是还原现代法式唯美的点睛之色，抑或是蓝灰色壁炉的基础用色，都能减少距离感，并为空间注入源源不断的温柔气质。古铜色装饰镜，不仅增强层次感，又可以带来舒适的视觉效果，和铜制水晶灯相辅相成，共同展示着空间不可忽视的高贵气质。

Formula | 方程式 ①

The personalized wallpaper matched with the calm and luxurious furniture ornaments can easily bring a modern sense of visual impact to the space. And the Norway blue brings a rhythm to the space, catering to the natural environment and presenting a tranquil flavor.

个性墙纸搭配沉稳不失华贵的家具摆件，可以轻易地为空间带来摩登的视觉冲击感，挪威蓝的点缀为空间带来了极美的蓝色韵律，迎合自然的环境，更有着宁静的意味。

Formula | 方程式 ②

The semicircle decorative cabinet in the foyer greatly eases the cramped feeling of the space and the personalized and exquisite furniture accessories make one feel that he/she is about to embark on a journey in the country.

　　玄关中半圆弧形装饰柜在空间中很大程度地缓和了局促感，个性精致的家具配饰摆件，好像在向我们慢慢呈现展示即将要踏上的国度之旅。

The natural color collocation is injected with the tension of life. The personalized green gold wallpaper is attached to the dark brown decorative wood, matched with the crown-like bed back, which is very luxurious and majestic. The bright and lively side cabinet painting and the orange blocks adjust the bedroom atmosphere, making the entire space as elegant and noble as a palace, with modern and natural flavor.

回归自然的配色，融入生命的张力，个性的墨绿镀金墙纸依附于深棕色装饰木作上，搭配如王冠般的床背，格外的华贵大气，明亮跳跃的边柜挂画和橙色墩子调节了卧室氛围，整个空间既有宫廷的优雅之贵，又具备现代自然的情调。

The blue vertical stripes on the wall weave a dream of ocean fairy land, giving children infinite reverie. All kinds of yellow cream color decorations are used to mediate the cold tone of the space, making people feel that they can follow the dolphins and begin a fantastic journey in the sunny ocean world.

将梦想照进现实,蓝色竖条纹交织似海洋国度般的梦境,给予孩子无限遐想的朦胧之美,空间的冷色调则通过各类奶黄色装饰摆件来进行调解,在阳光耀眼的世界里,跟着海豚,踏上去海洋世界的奇幻之旅。

On sunny days, light through the mottled shadows thickly sprinkles in the room with winding wisteria and blooming flowers. The green color collocation brings thoughts to a broader world where one could run in the green, ramble on the grasslands with deer and enjoy the freshness and leisure of nature.

天晴的时候,光芒穿过斑驳的树影,细细密密地洒在房间中,藤萝缠绕、花团锦簇、层林尽染、绿意盎然的色彩搭配将人的思绪带到更为广阔的自然世界。在绿地中奔跑,在草原中与鹿同行,将孩子带回自然,享受清新的悠然快乐。

英法风格 BRITISH & FRENCH STYLE

LUO YULI
罗玉立

项目名称：重庆湖山樾别墅

设计公司：深圳市则灵文化艺术有限公司

项目地点：重庆

项目面积：577m²

主要材料：实木、金属、布艺、水晶、皮革等

摄 影 师：曾康辉、黄书颖

Dwellers inside the Landscape
山水深处有人家

DESIGN CONCEPT ｜ 设计理念

The design of this case, whether in colors, lines or collocation, follows the delicate and rich French style, showing the owner's uncommon aesthetic taste. Exquisite diamond shape and complex classical dendritic images can be seen everywhere in the space. The softness of the soft fabric echoes with the charming luster of the metals, further enhancing the taste of space. The sapphire blue carpet decorated with European-style complex pattern unifies the front and the back of the living room. The modeling exquisite crystal chandeliers, wall lamps and table lamps highlight the huge classical European painting on the wall which echoes with the ancient fireplace at the opposite side, producing a roaring sense of time, bringing the cultural heritage color into the public space and conveying a bright, shiny and elegant space temperament.

本案设计无论是在色彩、线条还是搭配上，都遵循着细腻丰盈的法式风格，显示着主人不凡的审美品位。精致的菱形与繁复的古典枝蔓意象，在空间中随处可见，软包布料的柔软感与金属的迷人光泽相互衬托，进一步提升了空间的品味。饰有欧式繁复花纹的宝蓝色特制地毯，将空间开阔的客厅前后统一起来，造型精致的水晶吊灯、壁灯以及台灯，烘托着墙面上的巨幅古典欧式挂画，与对侧的古老壁炉遥相呼应，呼啸而过的时空感仿佛就要扑面而来，将文化的传承色彩融入到了公共空间，传达出光华、莹润、优雅的空间气质。

Elegant and warm, noble and free, just like the blue sky and the golden sun, they built the soul of this French mansion. This is a gift from the nature which purifies the madding crowd in the bustling city, and makes one feel like in "The Peach Garden". Quiet and harmonious, happy and pleased with oneself, this is the place where love resides.

优雅而热烈,高贵而自由,正如蔚蓝的天空与金色的骄阳,构建了这座法式大宅的灵魂。这份来自大自然的馈赠,净化了繁华都市里的车马喧嚣,让"桃花源"变得触手可及。安静和乐,怡然自得,这就是"爱"居住的地方。

观景阳台
起居厅
门厅
公卫
衣帽间
卫生间
过厅
老人房
前花院

品茶区　休闲区　书房会议区　酒吧区　庭院　保姆房　卫生间　卫生间　酒窖

庭院　品茶室　餐厅　厨房　男孩房　衣帽间　书房

Formula | 方程式 ①

The wise and elegant blue gives this mansion a fascinating French elegance. The gem-like noble blue makes the whole space distinguished yet smart while the golden and champagne decorations produce a sacred and rich tone.

睿智而优雅的蓝，赋予了这座大宅让人着迷的法式典雅气息。宝石般的高贵蓝调，使整个空间尊贵而不失灵动，金色与香槟色的点缀，混合出神圣而又丰饶的色彩。

Formula | 方程式 ②

With the sunken courtyard design, one can appreciate the charm of "The winding path leads to a secluded and quiet place" of traditional Chinese garden design. Compared with the flat layer structure in the residence, the hidden courtyard has a unique charm.

别墅采用下沉式的庭院设计，可以体味到中国传统园林设计讲究曲径通幽的意趣。跟一览无余的平层结构住宅相比，这种隐于内的庭院含而不露，别有一番韵味。

英法风格 BRITISH & FRENCH STYLE

GREEN TOWN DECORATION DESIGN GROUP
绿城家居设计团队

项目名称：绿城·珠海翠湖香山国际花园

设计公司：绿城家居

项目地点：广东珠海

项目面积：580m²

主要材料：胡桃木、金属、水晶、大理石、布艺等

Elegant Residence, Art Garden
优雅之居 艺术花园

DESIGN CONCEPT | 设计理念

Zhuhai Cuihu Xiangshan International Garden Villa Type B2, the entire area of which exudes a deep European royal atmosphere. The clean stone facade presents a grand and magnificent temperament in the sunshine. Indoor the rice white marble extends from the ground to the wall, showing the luxury and honor of the interior decoration. What jumps into the eyes is the shape and parquet of the entrance cabinet, making people feel like being in the 17th and 18th century French palace. The living room and the dining space are spacious with a three-people sofa and a double sofa symmetrically placed, making the whole space have a more complete streamline and strengthening the axis of symmetry of French style. The perfect use of the real and the virtual cleverly introduces the landscape outside into the room through the physical fireplace and the virtual view of the classic decorative mirror.

　　珠海翠湖香山国际花园 B2 户型别墅，整个小区的景观建筑就带有浓浓的欧式皇家气息，干净的全石材外立面，在阳光的照耀下让人感受到大气磅礴之势。室内米白色大理石从地面到墙面，无不展现出室内装修的奢华及尊贵。最直接映入眼帘的正是进门玄关柜的造型与拼花，让人置身于 17、18 世纪的法式宫殿中。客厅及餐厅空间宽敞，客厅三人沙发与双人沙发对称摆放，整个空间有了更完整的流线，同时也强化了法式风格的轴线对称感。实与虚的完美运用：实体壁炉加虚景古典的装饰镜巧妙地结合，将对面墙外景观引入室内。

Blue is the most representative color of the French style. In this spacious space, the eye-catching gem blue illuminates the dining room. Classical elements are interspersed in the fabric of furniture, accessories, carpet and curtains with different shades of blue, real or virtual, displaying the designer's color matching ability. There are two bedrooms on the first floor, respectively, a girl's room and a boy's room, the primary colors of both continue a blue tone, which are gray blue and lavender. The mature and fashionable room is for a girl who likes to paint, in which a mirror on the wooden furniture makes the space full of imagination. While the boy's room is the opposite, the main tone of which is warm and comfortable. The light blue makes people feel relaxed and warm in the afternoon sunshine. The unique study on the first floor is slightly tinged with domineering red, accompanied by a court-like book chair, showing the dignity of the host.

蓝色是法式最具代表性的颜色，在这个宽敞的空间中，宝石蓝夺人眼球，点亮了客餐厅。从家具、饰品，到地毯最后回到窗帘，面料中古典元素的穿插，不同灰度的蓝色有虚有实，展现了设计师的色彩调配能力。在一层有两个卧室，分别是女孩房与男孩房，都围绕着蓝色这一延续色，灰蓝色以及淡紫色。一个喜欢绘画的女孩，成熟又带些时尚，木质家具的镜面让空间充满了想象力。而男孩房则恰恰相反，温馨舒适则成为整体空间的主基调。淡淡的蓝色，在午后的阳光下让人感觉轻松温暖。一层别具一格的书房中，略带霸气的红色，配上宫廷般的书椅，显示出男主人的威严。

Formula｜方程式 ①

The use of color is essential. The overall color style of this case is unified and harmonious, with ingenious creation in subtleties. The case has a clever combination of classical and modern life philosophy, showing the amorous feelings of different tastes.

色彩的运用至关重要。整体色彩风格统一和谐，细微之处却别出心裁，巧妙地融合了古典与现代的生活理念，又展现出别有一番滋味的风情。

Formula | 方程式 ②

Paintings occupy a certain proportion in the living space. Choosing some appropriate works according to the overall style and atmosphere can enhance the artistic beauty of living space. This is not only an elegant and extravagant manor, but also an art garden in which one can appreciate beauty everywhere.

绘画作品在居室空间中占有一定的比例。从整体风格和氛围出发，搭配合适的作品能够提升生活空间的艺术美感。这不仅是一个洋溢着优雅贵气的庄园，也是一座能够处处欣赏美的艺术花园。

As the most private space, the second floor contains another study. Compared to the one on the first floor, it is much relaxed with a mini bar and a large decorative painting, showing the taste of the house-owners. The aisle is no short of interesting ornaments, with combination of decorative paintings and sculpture, which are humorous and witty. The master bedroom also employs blue, where the exquisite bed is matched with luxurious bedding, full of flavor. The blue color can be seen everywhere in this room: blue frames of the carpet, gem-blue fabric of the chairs, the blue satin fabric cushion, matched with the gray-blue stria wallpaper, making people involuntarily think of the scene in the film Marie Antoinette--the luxurious French palace with a perfect combination of furniture and decorations. The strong bar atmosphere is the most attractive part in this case, in which the droplight in the staircase illuminates only the piano area, creating a distinct contrast with the dark light and wallpaper. The personalized video room in the basement uses the contrasting colors of blue and red to light up the spirit. In addition, the yoga, chess and poker area in the basement are designed to blend with a French lifestyle.

二层，则是最为隐私的空间，楼梯之上又有一个书房，相比一楼的书房，这个书房要轻松得多，迷你吧搭配大型装饰画，显得男女主人个性情调十足，过道上不缺有趣的装饰，混搭的装饰画与雕塑的搭配，诙谐幽默。再来到主卧套房，同样运用了蓝色，精致的大床，加上奢华的床品，韵味十足。从地毯的蓝色套框到床位凳的宝蓝色面料再到蓝色缎面靠枕，层层递进，配上灰蓝的卷草纹墙纸，让人不由自主地想到《绝代艳后》中的场景，奢华的法式宫廷，家具和装饰完美结合。最后来到地下一层，浓浓的酒吧氛围是这套户型中最为吸引人之处，楼梯间的吊灯恰好只照亮了钢琴区，与灰暗的灯光与墙纸，呈现出鲜明的对比。地下室的影音室个性十足，蓝色红色的对比色运用，突然提起了精神，瑜伽、棋牌区都是以与法式生活相融合的方式去设计布置。

英法风格 BRITISH & FRENCH STYLE

Anna, Eenie

项目名称：滨江城市之星
设计公司：那依艺术设计（上海）有限公司
项目地点：浙江杭州
项目面积：390m²
主要材料：大理石、墙纸、石膏线条、木作、布艺、水晶、新西兰羊毛等
摄 影 师：陈兵

Luxurious Residence, Elegant Life
奢华居室 优雅生活

DESIGN CONCEPT | 设计理念

In home furnishing, this case adopts the French style, presenting the French romance. French Art Deco Style is most concentrated in the design of the furniture, and it highlights the symmetrical axis of the layout, which is magnificent, noble and elegant. In the handling of details, emphasis is put on carving, lines and exquisite workmanship. For example, the exquisite craving of the head and tail of the bed is solemn, generous and elegant, fully showing the owner's noble identity and status. In terms of the use of fabric, the addition of flower colors makes it no longer stuffy. In the whole, it abandons dark colors to avoid thickness, which is refreshing and impressive. Looking carefully, the exquisite carving and stylish and generous appearance exude a strong classical French atmosphere. The old white is very clean and eye-catching, simple yet elegant. In the carving process, emphasis is put on design details with gilt edged shape, which is very luxurious. Whether it is dining, or afternoon tea time, it can make people have a fantastic feeling of aristocratic life.

　　法式风格家居生活，体现法式浪漫之地，法国装饰艺术风格最集中体现在家具的设计方面，其特征在于布局上突出轴线的对称，恢宏的气势，高贵典雅。细节处理上注重雕花、线条，制作工艺精细考究。整体上的金碧辉煌不言而喻：床头的精细雕花庄重大方、典雅气派，充分彰显主人之高贵身份与地位。法式风格的特点是在面料上增加花色的衬托，不再古板。整体避开暗沉的色泽，摆脱沉重之气，令人眼前一亮。细看之下，拥有精美的雕刻工艺，时尚大气的外观造型，散发着浓郁的法式古典气息。造旧白的色泽很是干净，非常的夺目，在简单中不失高雅气息。在雕刻工艺方面，注重细节的设计，镶金边的造型更是奢华。无论是就餐，还是下午茶时光，都能让人有一种进入贵族生活圈的奇幻色彩。

Formula | 方程式 ①

The living room is dominated by a light color tone, with dark blue fabric dotted in the space to avoid monotony and dullness, which presents a sense of flexibility. The overall atmosphere is pure and elegant. This is an aristocratic house the designer built for the owner, and the occasional jumping colors add vitality to the living environment.

整体为浅色调的居室里，几抹深蓝色的布艺点缀其中，避免单调呆板，别有一番灵动之感。素净优雅的整体氛围，是设计师为屋主构建的贵族世家，偶尔的跳色为居室环境增添了几分活泼。

Formula | 方程式 ②

The exotic and elegant metal flowers exude a strong aristocratic atmosphere. Appropriately dotted in the space, they are not showy or overwhelming and the luxury they present accords with the entire living environment.

　　造型奇特优美的金属花卉，给人以浓厚的贵族气息。恰当地点缀其中，并不喧宾夺主，其显示出来的奢华，与整个居室环境协调一致。

英法风格 BRITISH & FRENCH STYLE

LUO YULI
罗玉立

项目名称：中航上城天玺大宅样板房

设计公司：深圳市则灵文化艺术有限公司

项目地点：福建晋江

项目面积：600m²

主要材料：实木、金属、布艺、水晶、皮革等

摄 影 师：张静

Luxurious and Elegant Fashion Style
华贵典雅 时尚气质

DESIGN CONCEPT | 设计理念

Zest Art respects the design style of the original hard decoration and combines the luxurious European style with the exquisite modern art style to create an impact of collision between the classic space form and modern furnishings, making the whole mansion flow a luxurious elegance and noble temperament of modern fashion. FENDI, a famous Italy luxury brand, is no doubt a synonym for luxury fashion. In the living room, a whole set of FENDI sofa and a round tea table are selected, exuding an intoxicating breeze of the Mediterranean Sea with the Rome gem inlaid in the mansion. The luxurious crystal chandeliers with rich level sense embody the visual focus of the living room while the yellow flowers on the round tea table endow the space with more color tension, exuding endless vitality.

则灵设计尊重原有硬装设计风格，将奢华欧式融入到精致现代的艺术风格之中，打造出古典空间形式与现代陈设相碰撞的冲击感，使整个大宅涌动着华贵典雅与时尚现代的尊贵气质。意大利著名奢侈品品牌芬迪FENDI，无疑是时尚奢华的代名词，来自芬迪FENDI的客厅全套沙发与组合圆几夹带着地中海彼岸醉人的海风，将罗马的瑰宝镶嵌在这座豪宅之中。富有层次感的水晶奢华吊灯，凝聚着客厅的视觉焦点。圆几上黄色插花，使空间色彩更具张力，同时也散发着不息的生命力。

Formula | 方程式 ①

The exquisite and stylish soft decoration not only reflects the quality of a comfortable and noble life, but reveals an ebullient and elegant temperament. The designer makes an ingenious combination of the visual impact and sense of space, calling the true desire of visitors.

精致时尚的软装陈设不仅体现了舒适高贵的生活品质,还彰显着灵动优雅的空间气质。视觉冲击感与空间感被设计师巧妙地糅合在一起,呼唤着参观者本真的欲望。

Formula | 方程式 ②

The noble and elegant purple, calm and wise midnight blue, bright and beautiful gold, graceful and cool gray, all these construct the soul of this house, making the whole space more colorful, gorgeous and attractive.

尊贵优雅的紫、沉稳睿智的深夜蓝、璀璨动人的金以及典雅冷静的灰,共同构建了这座豪宅的灵魂,使得整个空间在璀璨之外更有扣人心弦的华美。

MODERN STYLE
现代风格

现代风格 MODERN STYLE

SCD（香港）郑树芬设计事务所

项目名称：浙江·尚御府

主案设计师：郑树芬

软装设计师：杜恒、丁静

项目地点：浙江绍兴

项目面积：304m²

主要材料：索罗门米黄地台、瑞典橡木地板、墙纸、铜、大理石等

Elegance and Luxury Can Have its Own Unique Label
雅奢——可以有自我独特的标签

DESIGN CONCEPT | 设计理念

Never be controlled by a certain style or drift with the tide, have faith in the power of classic and culture and create works that can withstand the test of time, these are important parts of the concept of "elegance and luxury". In the case of "Noble Mansion, Zhejiang", the designer continues the low-key and calm restrained style of "elegance and luxury" to present exquisite techniques, in which the combination of classical and modern styles show the beauty of the coexistence of classic and luxury.

从不被风格左右，也不被"潮流"裹挟前进，坚信经典和文化力量，能经起时间考验的作品被称为雅奢主张的重要组成部分。"浙江·尚御府"这一户型中，延续雅奢低调沉稳的内敛手法，展现细腻的工艺，不被风格左右的古典和现代相结合呈现经典与奢华共存的美感。

As the most interesting element of space, materials such as classical copper, natural wood, smooth stones, hand-made woven leather bag with extreme modern elements, soft velvet of Oasis Italy brand furniture enrich spatial visual layering and texture experience. Fireplace is widely criticized in China, but if there is enough space, it would not only cater to people's habit of watching TV in the living room, but bring them an extremely wonderful feeling with warm flames in a cold winter.

材质是空间最好玩的元素,古典的铜、自然木、光滑的石、极具现代元素的手工真皮编织软包、Oasis意大利品牌家具丝绒的软等,丰富空间的视觉层次和质感体验。壁炉被中国人广为诟病,但如果有足够的空间,既满足客厅中国人看电视的习惯,在寒冷冬日,壁炉这一抹火焰带给我们的心理感受是极其美妙的。

In terms of soft decoration, the SCD creative team tend to apply multilevel texture, colors and original artistic works to create an artistic atmosphere for the whole space. Art is the most lively display of the space, which fully embodies the client's maximum acceptance and recognition of the designer's aesthetic. At the same time, the client's cultural and artistic taste can decide the quality of the project's result.

软装方面，SCD 创意团队喜欢注入多层次质感、颜色和艺术家原作来烘托整个空间的艺术氛围。艺术仍是空间最有生命力的展现，也充分体现了客人对设计师审美的最大接纳和认同，同时客人的文化艺术修养也能决定项目的实施完美程度。

As it is the designer's responsibility to enhance the spiritual connotation of the space, he makes deliberate consideration in the practicability of the space as well as the client's aesthetic taste. The sculpture works of artists Huang Yuan, Shen Zhou show a natural and simple taste; the hanging paintings of New York painter Carlos Ramirez, French painter Jean Francois Larrieu endow the space with color tension and artistic value, presenting the client's unique taste of living space.

在住宅项目上，兼顾实用的同时，赋予适合主人的审美情趣，提升空间的精神内涵是设计的责任，也体现客人的审美眼光。艺术家黄原、沈周来的雕塑作品，呈现一种自然简单的趣味；纽约画家Carlos Ramirez、法国画家Jean Francois Larrieu的艺术挂画，赋予空间色彩张力和艺术价值，并成为客人居住空间自我品味的独特标签。

Formula | 方程式 ①

Carefully-selected materials are used to create a new visual experience in the space, for example, the bronze pieces are cut in a modern style, endowing classic style with vitality of the new era, which is an interesting and innovative method.

材质的精挑细选打造空间新视觉体验，古铜片采用现代切割造型方式，将古典赋予了新的时代生命，也是有趣创新的玩法。

Formula | 方程式 ②

The interior of the house displays a number of artists' works, which not only show the owner's aesthetic interest and taste, but also highlight the inherent self-cultivation and cultural temperament of the space.

室内陈设着多位艺术家的作品，彰显出主人的审美情趣和自我品位，也凸显出空间内在涵养和文化气质。

As there is no standard language for design, we use comprehensive aesthetic to create a residential environment based on project orientation and client's preferences, while the concept of elegance and luxury is our pursuit and ideal in this case. We would like to express our special thanks to Mr. Zhang for his trust and support, and it is another ideal of Simon Chong design term to create a unique label for the clients in the concept of elegance and luxury.

设计没有标准语言，而我们用综合美学来诠释属于项目定位和客人喜好的住宅环境，用雅奢理念赋予我们对设计的一种追求和理想。特别感谢客人章先生的信任与支持，用雅奢打造属于客人独特的个性标签，是郑淑芬设计团队另一种理想。

现代风格 MODERN STYLE

WU WENLI, LU WEIYING
吴文粒、陆伟英

项目名称：深圳牧云溪谷悦溪郡32栋别墅样板房

室内设计：深圳市盘石室内设计有限公司、吴文粒设计事务所

陈设设计：深圳市蒲草陈设艺术设计有限公司

参与设计：陈东成、谢东泳、柯琼琼、罗楚希

项目地点：广东深圳

项目面积：1000m²

摄 影 师：张静、李林富

Toughness Lies in Entirety, Softness in Details
大处见刚 细部现柔

DESIGN CONCEPT | 设计理念

The soul of design lies in designer's ingenuity and unbridled pursuit of quality. During the repeated ultimate explorations, spaces are endowed with human emotions and temperature. It is a wisdom comes from life that the understanding of materials needs to accord their characteristics, yet a designer should forget his/her utilitarian idea and create the environment naturally. A good design should catch people's eyes and win their love and heart.

——Lu Weiying

"设计的灵魂在于匠心，在于对品质放肆的追求，在对极致反复的探索中，空间带上了人的情感和温度。那是一种源于生活的智慧，对于材料特质的理解，要合乎物性；而对于境界的缔造，则应忘记机心，自然而然。好的设计，让人目以投之、足以赶之、情以注之、心以神之。"

——陆伟英

Space design needs the designer's honesty and devotion. In this case, natural crystal jewelry and art paintings with manual temperature are displayed, revealing the designer's concentration like an urban craftsman.

空间设计需要匠人的守拙和专注。于是,在牧云溪谷悦溪郡别墅样板房设计中出现了天然水晶矿石饰品,出现了带着手工温度装置艺术挂画。每一个细节处,都透露着都市手工匠人般的专注。

The development of mansions reflects not only the city's economic prosperity, but the social pursuit of a better living environment as well. Starting from the addition of contemporary international prospective design thinking, to the discussion of relationship between urban and human, and the meaning that history and humanity have endowed on the space, this case seems to have proved that "man" is the most precious and irreplaceable part in a mansion no matter how luxurious it is.

豪宅的发展不仅反映着城市经济繁荣的程度,在某种意义上也承载了社会对于居住追求的梦想。从当代国际前瞻性设计思维的介入、到都市与人性关系的探讨、甚至是历史人文所赋予空间的含义,似乎都验证了再贵的豪宅,最珍贵、不能替代的,就只有一个字——"人"。

As human have feelings, beliefs and attitudes, the design should not be taken for granted. For the residents, "home" is not a concept, but a kind of experience and a place for dialogue between the true self and the world. The elites, promising, brave, able and responsible, no longer need those showy, tedious and conspicuous symbols to highlight their status. Confident in their own aesthetic, they do not go blindly with the streams. Instead, they are very particular about life quality and pay more attention to the spiritual world and the internal demand. Therefore, a private, warm, modern and international residence with strong personality caters to their rising vigor and strong vitality.

人有情怀、有信念、有态度，所以设计就没有理所当然。对居住的人而言，"家"不是一个概念，而是一种体验，一种"不忘初心"的与自己、与世界对话。有理想、有勇气、有作为、有担当的精英阶层，不再需要那些张扬的、繁琐的、炫耀式的符号来凸显自己的地位，他们对自己的审美充满自信，不盲目追随潮流，对生活品质极致考究，更加关注内在的精神世界及需求。私密、温暖、现代化、国际化，以及独特个性成为他们张扬朝气和强大的生命力诉求。

A successful design is like a poem, full of freedom and fantasy. The blend of contemporary art and luxury gives the visual organ a new experience and creates an irreplaceable texture. This case strikes to create a comfortable and practical space in addition to the creation of elite taste, making the concept of "home" in luxurious mansions become a pursuit of the comfortable modern life.

成功的设计,犹如一首诗歌,充满自由与幻想。当代艺术与奢华的融入,给予视角感官新的诠释,把设计做出不可替代的质感。强调精英品味的同时,着力打造空间的舒适性与实用性,使豪宅中"家"的概念真正回归到了对舒适的现代化生活的追求中。

The beauty of life is not blind, neither held by the past nor confused by the future. Based on contemporary aesthetics, this case re-examines the taste and creates toughness in entirety, softness in details, fully presenting the romance.

生活所需要的美不是盲目趋和,而是不被过去所挟持,不为未来所迷乱,立足于当代的审美,重新审视体面与品味。大处见刚,细部现柔,不着一笔,尽得风流。

Formula | 方程式 ①

The rugged and beautiful natural veins of the top-level stone create an impressive and charming scene together with mirrors, titanium metal and leather texture, presenting a fashionable, avant-garde, artistic imagination tension.

顶级石材粗犷且美丽的天然脉络与镜面、镀钛金属、皮革纹理,织就深刻动人的视觉印象,勾勒出时尚、前卫、艺术的想象张力。

Formula | 方程式 ②

The space color is constructed by blue of different levels, deep or light, extending from outside to inside, creating a dialogue between indoor device and material construction, which resonates with classic, gorgeous and modern.

空间色彩以或深或浅不同层次的蓝色自外而内导入,使得室内装置与材质建构的语汇之间,产生一种与经典、华丽、摩登共鸣的对话形式。

现代风格 MODERN STYLE

BENI YEUNG
杨铭斌

设计公司：硕瀚创研
陈设艺术：东西无印
项目地点：广东佛山
项目面积：630m²
主要材料：肌理漆、墙布、地毯、烤漆板、大理石等

Splendid Color
流光溢彩

DESIGN CONCEPT | 设计理念

You can spend a few hours here, silently watching the river in front of you and letting thoughts slowly flow…
This house is endowed with natural beauty. Planning of the house's functions has begun since design of the architecture. Therefore, the designer does not drastically re-construct the indoor layout but deliberates on details so as to make functions of the interior space get close to be perfect.

你可以花几个小时呆在这里，静静地看着眼前的江景，让思绪慢慢流淌……
这栋房子就是拥有得天独厚的天然美景。房子的功能规划应该从建筑设计时就已考量了，因此设计师并没有大动作地把室内建筑重新分拆，而是从细节上考究，让室内空间功能接近完美状态。

On the other hand, the designer deliberates on the color collocation in the space. The elevator is at the center of the house around which the stairs extending upwards. This case selects geometric pattern on the elevator's wall in order to set the atmosphere of the entire house and render the continuation and transition in the space. You can see these colorful geometric tiling pictures whether you turn left or right, look upwards or downwards.

另一方面，设计在空间里的色彩搭配研究，房子中心位置是电梯，楼梯围绕着电梯拾级而上，通过选取几何图形色块的图案用在电梯墙壁上，目的为了锁定整个房子里的氛围，更表现出空间上的延续与过渡。无论你向左或向右，还是抬头或低头，都能看到这些色彩斑斓的几何图形拼砌画面。

The designer takes down the partition between the living room and the staircase to introduce light into the underground staircase and make lighting and ventilation convection achieve the best effect. After the demand for functions has been met, the form is naturally created, enriching the layers between the living room, staircase and dining space.

设计师将客厅与梯间的隔墙拆掉，让光线引进通往地下一层的梯间，使其采光性与通风对流达到最佳效果。当满足功能需求后，其形式自然而成，使客厅、楼梯、餐厅之间的空间关系层次更显丰富。

Formula | 方程式 ①

With white as its main color, this house overlaps the colors in the hope to highlight the artistic taste and touch the owner's heart through "form" and "color".

整个房子的格调以白色调性为主体，通过色彩来叠加，希望通过"形"与"色"突出艺术品位，让"形"与"色"触动生活其中的主人心灵。

Formula｜方程式 ②

This case pays attention to the residents' sense of belonging in the design. Selecting materials according to their "form" and "color", the designer breaks through doctrines so as to make users interact with the design.

设计师在乎生命体在设计中的归属感，取材于"形"与"色"，呈现概念，摒弃主义，让用户和设计本身互动。

现代风格 MODERN STYLE

LU YINA, ZHU ZHIYI DESIGN TEAM
卢燚娜、朱芷谊团队

项目名称：七彩云南·古滇王国文化旅游名城—湖景林菀400户

设计公司：奥迅设计|奥妙陈设

项目地点：云南昆明

项目面积：625m²

主要材料：不锈钢、大理石、水晶、花艺等

In Search of Lost Time, Low-key Luxury
追忆似水 奢华低调

DESIGN CONCEPT | 设计理念

In traditional sense, the word "luxury" is often associated with complexity, dignity and beauty, and regarded to pay attention to ornamental value yet lack of practicability. However, modern luxury usually emphasizes on low-key luxury whose design techniques are simple yet not casual. The seemingly simple appearance often reflects a hidden aristocratic temperament, presents exquisite soft decoration in details and contains unexpected functions and elements of technology.

In the entrance, the Cubist paintings with mosaic geometric figures laid the decorative style of the whole space. The retro symmetrical round yellow lamps on the wall with bright wood finishes are well matched with the three-dimensional gold which embellishes the ceiling. The entrance provides a free access to the corridor of the living room, giving people a bright and casual impression. The living room uses the traditional open space (Great Room) technique to suit the bar, heighten the space in the core position, and define the function of different spaces through sofas. The use of stunning retro chandelier, comfortable high-level fabric, simple and stylish carpet, as well as furnishings favored by the owners creates a level sense of space.

传统观念中，"奢华"一词往往和繁复、尊贵以及精美联系在一起，注重观赏价值，缺乏实用性。而现代奢华，通常注重低调的奢华，设计手法简洁、不流于随意。看似朴素的外表，时常折射出隐藏的贵族气质，以精致的软装细节体现，还浓缩着意想不到的功能与科技元素。

玄关，"玄之又玄，众妙之门"，一副几何图形拼接的立体主义画作，奠定了整个空间的装饰主义风格。两旁复古对称的圆型黄灯与点缀天花的立体金甚是搭配，联结着亮光木饰面墙身，自由通向客厅的廊道，予人以敞亮、随性的"第一印象"。客厅运用传统的开放式空间（Great Room）的手法与吧台相适应，在核心位置挑高优势，依靠沙发来定义区分机能。利用复古矜贵的吊灯、舒服高级的布艺、简约时尚的地毯，带着主人喜好的陈设品，打造出空间的层次感。

The dining room is lit up by a huge ring-shaped crystal chandelier. With the embellishment of classic coffee leather chairs, the space shows a low-key luxurious and majestic atmosphere. Beside the dining table, the collocation of furnishings presents a strong contrast of blue and orange, creating a visual perception with both a surreal and retro flavor, showing the intersection of modern and tradition and giving the space a new life.

餐厅中巨大的圆环形水晶吊灯倾泻而下,点亮了用餐空间,配以经典咖的真皮餐椅为点缀,呈现低奢、大气之风。用餐旁,以蓝与橙为主的强烈对比的陈设搭配,碰撞出兼具超现实与复古风味的视觉感受,表现出现代与传统的交错,寄予空间一种新的生命。

The lounge also uses the "Open Floor Plan", which is linked with a mahjong table. The sliding door leading to the red wine house adopts both functional and aesthetic design, creating a rich visual effect and providing more possibilities for interaction. After meals, the owner could play mahjong with his intimate friends or have free talks over wine.

休息厅，再次使用了"开放式空间"（Open Floor Plan）的设计，与麻将桌联结在一起；通向红酒屋的推拉门，兼具功能与美感的设计，营造出丰富的视觉效果，为互动创造了更多可能性。饭后小憩，或与三五知己一展"技艺"，抑或是在酒过微醺之时，随性畅谈。

In the master bedroom, the modern bedside lamp softens the thick and eye-catching orange of the whole space. The minimalist white of the carpet echoes with the bed furnishings, together with the background sailing painting they present the master's indomitable and dauntless exploration spirit and free personality. In the children's room, the simple yellow brown carpet brightens up the deep wooden floor and adapts to the brightness of the room. The whole space is dominated by the sunny and youthful blue and orange and decorated with a surreal and three-dimensional "horse" art painting, symbolizing the "manliness" and "male power" of the owner.

主卧中一盏现代床头吊灯的陈设，低调了整个空间浓烈、醒目的橘红，地毯上简约的白与床上家私相呼应，再配以背景的帆船油画，透漏了主人一往无前、无所畏惧的探索精神、自由洒脱的性格。孩子房中简约的黄棕色地毯提亮了原木地板的深沉，并与房内的灯光亮度相适应；空间以阳光、青春的蓝与橙为基础，缀以"马"为主体兼具超现实与立体的艺术画作，寓意屋主为"阳刚""雄性力量"的象征。

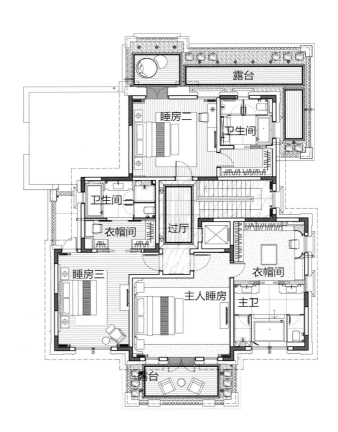

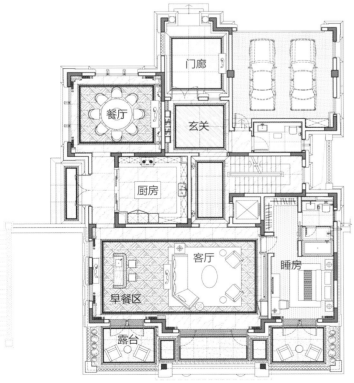

Formula | 方程式 ①

The designers follow the modern classical elements and combine the modern urban design techniques to create a bright and strong color contrast with retro furniture and decorations, presenting a luxurious space with equal attention to fashion and practicality.

设计师承袭摩登经典元素，融合现代都市设计手法，借以复古家具、饰品的陈设，明亮、强烈的色彩对比，呈现一个时尚与实用并重的奢华空间。

Formula | 方程式 ②

The case uses a city building painting as the main background of the modern industrial lights and bar, rendering the fast tempo of urbanization and modernization. With a cup of Whisky, a scene usually appears in the Hollywood movie is presented.

摩登现代的工业灯与吧台主要背景——城市建筑画作，共同渲染了一种城市化、现代化进程的快节奏感，再来一杯 Whisky，即刻变身 Hollywood 电影经常出现的一幕。

ZHU JUNXIANG
朱俊翔

项目名称：深圳华侨城-天鹅堡一期

设计公司：维塔设计

项目地点：广东深圳

项目面积：220m²

主要材料：皮革、大理石、挂画、不锈钢等

摄 影 师：江南摄影

The Integration of Elegance, Luxury and Comfort; Function and Beauty

优雅奢适 功能与美的融合

DESIGN CONCEPT | 设计理念

With "low-key nobleness" as the artistic creation spirit, this case is not artificial, exaggerated, or noisy, but it expresses its own fashionable attitude. As Ando Tadao said, "a luxurious home must have a quiet feeling and touch the soul."

The entire space has a calm and quiet color scheme, uses wise and calm palm color, noble and elegant gray as the basic tones to match with the natural and luxurious rose gold, Hermes logo classic orange. Besides, an addition of a neutral tone to the place creates infinite comfort. The furniture material and style reflect the personality and pursuit of the upstart family. A variety of furniture and ornaments are cleverly and flexibly placed in the space with unique metal lighting, painting and art decoration, making the artistic inspiration flow everywhere.

以"低调的高贵"为艺术创作精神，不造作、不浮夸、不喧嚷，以此表达空间自身的时髦态度。正如安藤忠雄所说："奢华的家要有安静的感觉，触动心灵深处。"

整个空间色彩沉稳而安静，以睿智冷静的棕榈色、高贵典雅的灰为色彩基调，搭配自然华贵的玫瑰金、爱马仕标志的经典橙，以及中性的色调在比例、情绪和故事间平衡出了无限的舒适；家具材质和款式方面体现了新贵一族的个性与追求，各种家具、饰品在空间中生动巧妙地并置，而别具一格的金属灯饰、挂画和艺术装饰，无不流动着艺术的灵感。

The entrance integrates strong visual symbols to the space, such as the contemporary artistic hanging paintings, decorations and furniture, forming a temperament that is close to life yet obviously higher than life. Complex decorations are abandoned while the most concise and smooth lines are kept in the living room with stylish advanced custom furniture and the spacious space complementing each other, fully presenting the comfort of life. Quiet gray, simple khaki, elegant light brown and charming metal lines enhance the space temperament. Combined with the handmade leather fabric of the private custom furniture as well as contemporary ornaments and metal texture, a unique space temperament is created, showing a unique aesthetic and tolerance.

玄关以当代艺术气息的挂画、装饰、家具等形成强烈的视觉符号转换到空间中，形成接近生活又明显高于生活的气质。

客厅中，时尚大气的高级定制家具与宽敞的空间相得益彰，去除繁杂的装饰留下最精简干练的线条，极尽表现生活中的舒适度和舒适性。静谧的灰，素简的卡其色、优雅的浅棕与散发迷人魅力的金属线条提升空间气质。结合私人定制家具的手工感皮革面料，以及当代饰品与金属质感，形成独特空间气质，隐喻独到的审美和气度。

The straight lines in the dining room open the space dimension like a flexible flashlight. The recombination of the models and the inheritance of quality make the space calm, modern and elegant. In the master bedroom, the bright yellow bed runner, unique furnishings and decorative painting embellish the space, exuding an artistic flavor and the pleasure of life. In the bathroom, the simple color on the ground creates a visual impact, bold, brave and harmonious.

餐厅笔直的线条犹如一道灵动的光束拉开空间的维度，样式的重新组合与品质承袭，让空间集沉稳、现代、雅趣于一体；并生动巧妙地汇集在一起。主卧明亮的黄色床旗和别具特色的陈设、装饰画点缀在空间中，链接起了空间的艺术性，也散发着随心所欲的生活乐趣。主卫地面简单的色彩视觉冲击，大胆、勇敢而和谐。

Formula｜方程式 ①

The post modernism style is adopted in this design with color and furnishings to achieve the specific functions and unique charm of different regions, showing a rich aesthetic of the space. The use of a large number of intermediate colors integrates the space whose main tones are gray and palm and collocated colors are mainly khaki and beige. Together with a small area of color collisions and collocation of Chinese and Western elements, the overall environment is calm and elegant yet not short of changes.

设计主要以后现代主义风格为基础环境，以色彩和陈设来实现不同区域的特定功能和独特韵味，呈现空间丰沛的美学力量。运用大量的中间色系让空间融合起来，以灰色和棕榈色为主色调，搭配色调以卡其、米黄色为主，小面积的色彩碰撞，中西方元素搭配，令整体环境沉稳优雅而又呈现变化。

Formula | 方程式 ②

With modern minimalist style, flamboyant color collocation and novel materials, it has become an art symbol of the current fashion design. In the design of private space, exquisite and luxurious details of the decoration present a high-level style of the natural form. The balance of function and beauty is a perfect expression of a comfortable, pleasant and peaceful environment, secretly reflecting a luxurious temperament.

通过现代极简的造型、艳丽夺目的色彩以及新颖的材料搭配,成为当下时尚设计的艺术符号。在私宅空间设计中,精美奢华的细节装饰表露的却是自然形式的高级风格化。功能与美的平衡将一个舒适、宜人、安宁的环境表达得恰到好处,体现暗自华贵的气质。

现代风格 MODERN STYLE

Han Song
韩松

项目名称：长沙恒伟西雅韵项目A户型

设计公司：深圳市昊泽空间设计有限公司

参与设计：吴海蓉、陈晨艺、王娜、陈文雅、马超

项目地点：湖南长沙

项目面积：350m²

The Storyteller
讲故事的人

DESIGN CONCEPT | 设计理念

Urban life is easy to produce sentimental natural feelings. The city dwellers see nature as a garden, or a scenery outside the window, or a free stage.

It is in this natural situation that we experience beauty. In essence, the encounter of beauty is unpredictable. For example, in the calm and tranquil sea, the water color turns from earth gray into sapphire blue; flowers grow under a huge rock which rolled down in an avalanche; the moon rises in a dilapidated town. Looking at more examples in daily life, no matter how we encounter beauty, beauty is always an exception, and that's why we are touched by beauty.

Art is not an imitation of nature but an imitation of creation. Sometimes art suggests another world, but sometimes it just amplifies, confirms or publicizes the simple hopes of nature. The art systematically responds to what nature allows us to catch glimpses occasionally. In a word, art is designed to translate the potential cognition into perpetual cognition.

　　城市生活容易让人萌生感伤的自然情怀。城市人看自然是花园，或是窗外的风景，或是自由的舞台。

　　正是在这样自然情境里，我们遭遇美。这遭遇究其本质是不期然的、无法预料的。风平浪静，海水从土灰变成宝蓝；雪崩后滚落的巨石下长出小花；破败小镇的上空升起月亮。反观更多日常的例子，无论我们是如何遭遇美，美始终是个例外，这正是美打动我们的原因。

　　艺术不是模仿自然，艺术模仿创造，有时候艺术举荐另一个世界，有时候只是放大、证实、宣传自然所提供的简单希望。艺术有条理地回应自然允许我们所偶尔瞥见的。艺术旨在将这个潜在的认知转换成永恒的认知。

Ultimately, the story does not depend on what it says nor the so-called complex -things that in our cultural paranoia and be projected into the world. It does not depend on any fixed ideas or habits, instead, it depends on its stride across the space. In these spaces, there is the meaning that stories endow on events, most of which come from the common aspiration of characters and readers in the story.

归根结底，故事不依赖于它所述说的，不依赖于我们将自己文化妄想症的某些东西投射到世界上，即所谓的情节。故事不依赖于任何思想或者习惯的固定保留剧目：故事取决于它跨越空间的步伐。在这些空间里，存在着故事赋予事件的意义，这种意义绝大部分来自故事中的人物和读者之间共同的渴望。

Space is unique and meaningful because of the story and the people who created it, and the case is intended to tell the story of space in a design language. The picture in the living room condenses the indoor focus, of which the color integrates harmoniously with that of the room. At the same time, the "from the close to the distant" perspective of the painting produces a space extension effect for the living room. With the embellishment of the crystal chandelier and fragrant flowers, the dining room exudes a romantic and delicate flavor. The master bedroom is quite spacious with an open study, exuding a scholarly atmosphere and creating a quiet atmosphere of the space. When the night is falling, it is very quiet in the space where one can have a peaceful rest after work.

空间因有故事和创造故事的人而变得独特而有意义，本案则试图以设计语言去诉说空间故事。客厅以一副挂画凝聚室内焦点，画作颜色与客厅色彩和谐相融，同时挂画由近及远的角度也让客厅产生空间延伸的效果。餐厅在水晶吊灯和芬香的插花映衬下，显示出浪漫的精致气息。主卧空间较大并设计了一个开放式的书房，融合了书香气息，创造了空间的静谧氛围。暮色四合，夜深人静，伏案疲劳之时便能在此安然入眠。

Formula | 方程式 ①

The interior decoration is dominated by neutral and cold tones, showing the low-key texture of the space and letting the colors full of harmonious unity. The calm wood color, low-key gray and soft beige present the best color collocation of the space.

室内装饰以中性色调和冷色调为主,显示出空间低调质感,也让色彩上充满和谐的统一。木色的沉稳、灰色的低调、米色的柔和等,赋予了空间最好的色彩搭配方式。

Formula | 方程式 ②

The design of the boy's room fully demonstrates the child's young and lively character. The wallpaper of vertical white, blue and red strips is dynamic and echoes with the red and blue bed furnishings, while the decoration with music elements reflects the boy's hobby and shows his youthful personality.

男孩房的设计充分彰显出孩子年轻和活泼的特质,竖条的白、蓝红色相间的壁纸富有动感气息,床品也以红蓝相配与之呼应,音乐元素的装饰体现出男孩兴趣爱好和标榜着的青春个性。

现代风格 MODERN STYLE

HARMONY WORLD CONSULTANT & DESIGN
HWCD设计

项目名称：金臣别墅A1

项目地点：上海

项目面积：677m²

主要材料：壁纸、石雕、雀眼木、云纹玉石、丝质织物、木质等

Modern Interpretation of Fashion Orient
时尚东方的现代演绎

DESIGN CONCEPT | 设计理念

This case is characterized by an elegant and fresh overall tone and exquisite natural elements. The interior design interprets the Oriental people's pursuit of elegance, lightness and freshness in an implicit and aesthetic way. The traditional Chinese residence is partial to "elegance", in which the color collocation with relatively low brightness gives people a sensation of tranquility and harmony. Focusing on this criterion, the design team selected several kinds of standard colors from both the English royal colors and the Chinese colors to form the interior color system, in which the brightness of the English colors and the light elegance of the Chinese colors bring out the best in each other.

淡雅清逸的整体色调、细腻精致的自然元素是本户型的特色。室内整体设计以含蓄唯美的笔触演绎了东方人心性所追求的雅致、轻快和灵活。传统中国居室用色讲究"淡雅"，纯度、明度都较低的颜色令人感到宁静和谐，这也是东方人最热衷的色彩搭配。设计团队以此为准则，从英国遗产色和中国式颜色中选取了若干种标准色构成室内色彩系统，英式色彩的明亮和中式色彩的淡雅相得益彰。

The living room is dominated by light colors, with tooth yellow jade on the wall which serves both as a background wall and a decorative fireplace. The space is dotted with birds-eye wood rich in exquisite and natural texture. The sofas and pillows adopt crimson and blue green respectively, which represent traditional Chinese painting colors, exuding a subtle Oriental flavor. The decorative fabrics are interspersed with the English colors: the elegant Dorchester pink, lace white and bright marigold, making the space more rich and vivid.

客厅整体以缟色为底,墙面上牙黄色的玉石兼具背景墙和壁炉的功能,缃色的雀眼木出现在空间各处,富有细腻自然的肌理。沙发与靠枕分别选取了代表中国传统丹青色彩的绯红及雪青色,散发出含蓄的东方气韵。

而英式色彩则更多地被穿插运用于装饰面料中,淡雅的多切斯特粉、格蕾丝白、明亮的金盏花色,使空间层次更丰富、生动。

Formula | 方程式 ①

The fusion of decoration and nature is essential for suburban houses. The designers extract a large number of basic elements from mountains, water, trees and create a series of indoor decorative details through abstract transformation and London handicraft carving. Ginkgo decorative painting on the wall of the living room, flowers and birds wallpaper on the master bedroom wall, corrugated pattern of marine theme on the cushion, all these create a rich natural flavor.

装饰与大自然的融合，是郊区大宅内不可或缺的组成部分。设计师从自然界的山、水、树林当中萃取了大量基本元素，通过抽象转化和伦敦手工艺的雕琢，形成一系列室内装饰细节。客厅墙面上银杏装饰画，主卧墙面上的花鸟壁纸，靠枕上源自海洋主题的波纹图案，营造出浓郁的自然气息。

Formula | 方程式 ③

The charm of colors is used to highlight the theme. In the master bedroom, the ivory yellow wallpaper with London hand-painted Wisteria flower pattern makes the whole space full of soft life atmosphere. Brown stones with natural texture are applied to the washbasin, the ground and the separate tub carved by a whole stone in the master bathroom, which has a unique style highlighted by metal lines. The basement adopts the ancient and heavy deep red as the background color of the whole space with silk wallpaper and solemn wooden lights, creating a refined taste full of Oriental atmosphere.

运用色彩的魅力，突出主题。主卧设计以牙黄色基底配伦敦纯手工绘制的紫藤花卉图案壁纸，使整个空间充满柔和的生机。主卫内的台盆、地面及以整块原石雕刻的独立式浴缸采用统一的褐色自然纹理石材，在金属线条的勾勒下别具一格。地下室选取了古厚的绾色作为整个空间的背景，绢帛质感的丝质壁纸、沉稳的木作展架灯火，营造出充满东方气质的文房雅趣。

Formula | 方程式 ②

This case has a good grasp of details, perfectly showing the quality of space. The designer adopts the most representative traditional oriental elements and London custom techniques and materials and introduces New London life style, showing the highest quality of home furnishing. The fireplace background wall made of traditional jade in the living room, the metal partition evolved from Chinese window edge, the embroidery flower-and-bird patterns on the wall of the bedroom, the tea table with metal frames in the study, all these are epitomes of Oriental Art, bringing a different and refreshing kind of interest.

把握细节，完美展现空间品质。设计师运用了极具代表性的东方传统元素和伦敦定制工艺与材料，引入了新伦敦生活风尚的同时，展现了家居生活的最高品质。客厅内传统玉石制作的壁炉背景墙，由中式窗棂演变而来的金属隔断，以传统苏绣方式呈现的花鸟图案床背、书房内案几造型的金属边桌，随处可见优雅的东方艺术缩影，带来令人耳目一新的别样趣味。

现代风格 MODERN STYLE

SCD
（香港）郑树芬设计事务所

项目名称：海南·万科森林五期别墅样板间

主案设计师：郑树芬、徐圣凯

软装设计师：杜恒、丁静

项目地点：海南三亚

项目面积：220m²

Forest Holiday: Great Beauty Lies in Simplicity
森林度假风：简约之中存有大美

DESIGN CONCEPT ｜ 设计理念

Relying on the original Ecological Forest Park, this project is surrounded by hills and streams with the hills at the back and the streams at the fore. It enjoys scarce resources of the original ecology and creates a one-stop family resort for modern city dwellers. As the project meets the requirements of tourist resort, it is a standard exquisite holiday villa project. There are three storeys in this villa, including the basement. On the first floor, there are functional spaces such as living room, dining room, kitchen and bathroom, in which the designers cleverly used white color widely. Forest is always the best theme of nature. With a bed, a chair and a book in the forest, one can escape from the bustle and hustle of the city. In this villa, the reeds under the setting sun, the brisk rabbit, the dandelions casually planted in a bottle and the unique cactus sculptures combine the best picture of a forest holiday. In the evening, when the glass table lamp lights on, one feels like being in a reed forest illuminated by a full moon.

海南三亚万科森林度假公园项目依托原有的生态森林公园，沿山丛溪流依山而建，坐山朝南，背山面水，四周群山环绕，享有原生态的稀缺资源，为现代都市人创建一站式家庭度假乐园。该项目以旅游度假定位为要求，是精装交楼标准的度假别墅项目。含地下室共三层的精致小别墅，一层含客厅、餐厅、厨房和洗手间为主要功能空间，设计师巧妙地大面积采用了白色。

森林，永远是自然最好的主题，在森林里的一张床，一把椅，一本书，一洗城市的喧嚣。夕阳下的芦苇草，神气活现的兔先生，随意植入瓶中的蒲公英，独一无二的艺术家作品仙人掌雕塑，组合成一幅森林度假最好的画面，夜晚点亮玻璃台灯，就像圆月照亮的芦苇林。

Formula ｜ 方程式 ①

A light pure white space can wash away irritability and relax the mind. Soft decoration of this case prefers casualness to pretensions, enhancing significantly the residents' holiday feeling.

轻盈的纯白色空间能拂去烦躁，让人的思绪得以放松和平静，软装讲究随意性而不做作，让居住者的度假感受得到更明显的提升。

Formula ｜ 方程式 ②

Great beauty lies in such details as distressed iron, retro wood and paintings with cultural ambience. These details make the space exude a cultural temperament and family warmth. This is what a home should look like, which embodies what SCD has advocated: the holiday exquisite show flat should return to nature and simpleness.

做旧的铁艺，复古的木头，带有文化笔触的挂画，用细节在简约之中表现大美，让空间流露出文化气质和家的温情，这是一个家该有的样子，也是 SCD 倡导度假精装样板房回归简单与自然。

Simple and not dazzling, reserved as a virgin, quiet and graceful, here is the home of soul rather than the so-called cold mansion. Warm sunshine sheds on one corner of the stairs which the designers cleverly decorate into a children's paradise of reading in which the abstract fish patterns on the carpet, the hanging picture of snowy grassland, the bunny, Baymax and tiny sheep placed in the bookcase serve as the partners of angel.

朴拙而不耀眼，含蓄如处子般静谧婉约，置身其中，这里就是心灵的家，不是人们口中冰冷的豪宅。阳光的暖，洒落在楼梯的一角，设计师巧妙地把它装点成儿童的读书乐园，地毯的抽象鱼纹图案，冰雪草原的挂画，静摆在书柜的兔宝宝、大白和小绵羊，成为天使的伙伴，这应该是家的一部分。

Children's room is casual and interesting with a bark clock, elf sculpture ornaments and tiny robots under the bright lamp, which are the scenes most familiar to the children. The basement is used as a supporting space in which careful consideration has been made in lighting and function. The designers create a level sense of space through a sunken courtyard and a patio to maximumly introduce sunlight into the room and create a spacious and bright leisure area with green landscape. The lemon yellow kayaking on the wall makes one feel that he can go to the beach with it at any time and enjoy the beauty of nature and life.

儿童房随意而有趣，树皮做的挂钟，独一无二的小妖雕塑摆件，明黄和明蓝的对比，明月般台灯下酷酷的机器人，都是孩子们最亲切熟悉的场景，也是一个家快乐的画面。负一层作为配套空间，在采光和功能上做了细致考量，通过下沉庭院、天井来打开空间的层次感，将阳光最大限度地引伸到室内，塑造宽敞明亮的休闲区域，绿植景观，柠檬黄的皮筏艇，随时背起走向海边的洒脱，感受自然与生命的纯美。

NEO-CHINESE STYLE
新中式风格

新中式风格 NEO-CHINESE STYLE

ZHANG LI
张力

项目名称：中海宁波九塘酌月样板房

设计公司：飞视设计

参与设计：陈邵云、赵静、王佳元

项目地点：浙江宁波

项目面积：128m²

主要材料：兰金莎大理石、实木复合地板、手绘壁纸、蚀刻面古铜不锈钢、乳胶漆等

摄 影 师：张静

Mountain House
居山院

DESIGN CONCEPT ｜ 设计理念

On a winding path leading to a secluded quiet place, enjoy a touch of elegant red antefix and green tiles; one feels like walking in a road to the childhood's home. When opening the door, the uninvited sun is shining brightly. Deep in the fragrance of plum blossom is the home to the Chinese, where the drizzle is falling on the grass in the courtyard, and also tapping on the homesick heart.

曲径通幽处，采撷一枚朱檐青瓦的淡雅，偶遇一段儿时的回家路。推门而入，是不请自来的斜阳，还有梅香深处中国人的归属，细雨敲打着一院碧水青草，也敲打在想家的心头。

Located in Dongqian Lake tourism resort, this case is against the mountains in the southwest and faces the lake in the northeast with an open lake scenery. Endowed with unique lake scenery and advantaged geographical location, this case adheres to the design concept of "mountain, water, home", creating a villa group with lake scenery and picturesque landscape.

本案坐落于东钱湖旅游度假区，西南依山，东北面湖，湖面开阔，景色绝佳。拥有不可复制的一线湖景，地理位置得天独厚。秉承"一山一水一世家"的设计理念，打造出"一线临湖，领袖山水，别墅群落"。

Designers try to find a kind of modern emotions from the Chinese courtyard to create the relation between man and space, giving people unique impression and experience and producing a sense of closeness. The relation between the interior and the exterior, between architecture and nature, tradition and the contemporary, is formed to enhance people's environmental experience. Designers hope to put aside all forms and labels so as to integrate traditional culture into modern culture. Through exquisite refinement of traditional elements, the beauty of nature and the beauty of humanity are completely integrated, creating a kind of tranquility in the bustling city.

设计师尝试从中式院落中寻找一种属于现代的情感，将人和空间创造关系，给予人们的观感和体验是独特的，产生亲近感受。内与外的关系、建筑与自然的关系、传统与当代的关系，而这系列性关系却是围绕着提升人们的环境体验而展开的。设计师希望抛开一切形式和标签的表象，将传统文化融入到当代设计中，传统元素的精湛提炼，将自然之美与人文之美完全融合，在繁华的都市下寻找一种宁静。

Formula ｜ 方程式 ①

This case makes good use of geographical advantages to introduce the natural scenery into the interior. The hard top roof with five ridges and two slopes of the courtyard is extended into the interior, achieving a balance between the traditional and the modern. The peaceful and simple natural beauty stimulates the new senses under the collision of wood and stone, presenting the slight movement in the space and sketching out a leisurely and quiet landscape.

　　善于利用地理位置的优越性，将自然之景引入室内。将院落的硬山顶建筑形式延伸至室内，五脊二坡，在传统和现代之间，寻得二者平衡。安宁朴素的自然美感在木与石材的碰撞下激发新的视觉感受，呈现出空间中细微的感动，笔墨深浅，寂寥无声，勾勒出悠闲的山水意境。

Formula | 方程式 ②

Local cultural characteristics are highlighted in this case. The "quietness" and "cleanness" of Dongqian Lake lead people to a cloistered life of "living in a deep forest unknown to all, with only the moon shining in the sky". This case adopts an open courtyard to combine the outside with the inside, restoring the essence of space and recovering the original and simple beauty.

抓住地方性，突出当地人文特色。东钱湖的"静"与"净"，让人沉浸于"深林人不知，明月来相照"的小隐生活，结合开放式庭院，内外合一，回到空间的本质，创造出一种返璞归真的美。

新中式风格 NEO-CHINESE STYLE

WANG XIAOGEN
王小根

项目名称：柳州悦景台

设计公司：北京根尚国际空间设计有限公司

项目地点：广西柳州

项目面积：400m²

主要材料：木饰面、石材、大理石等

摄影师：史云峰

Melodious Ancient Rhyme
悠悠古韵

DESIGN CONCEPT │ 设计理念

This case adopts the modern Chinese style with plenty of Chinese elements such as landscape paintings, porcelain, paper fan, pipa, tea table which add an ancient flavor and unique interest to this modern space. Modern furniture such as sofa and modern crafts create a comfortable and warm home atmosphere for the space. The coexistence of conflict and harmony makes the space full of "accidents" and "surprises" and the fusion of tradition and modernity makes the home atmosphere more harmonious and natural.

本案为中式现代风格，里面的中式元素比比皆是，山水画、瓷器、纸扇、琵琶、茶桌等等，这些装饰品为这个现代化的空间中增添了一抹古韵和别样的趣味。现代化的家居，如沙发、现代工艺品等为居室营造了舒适亲切的居家氛围。冲突与和谐并存，使得居室处处有"意外"和"惊喜"，传统与现代的相互交融，使得居室氛围融洽自然。

Formula | 方程式 ①

The square lattice ceiling reflects the ancient Chinese theory of "round sky and square earth". The use of large areas of squares makes the space look more transparent and spacious. It agrees with the overall deep tone, making the space more steady and majestic.

　　方格子的天花板契合古人天方地圆的遐想，大面积的方格子运用，让空间显得更通透旷达，配合整体的深色调，更显沉稳大气。

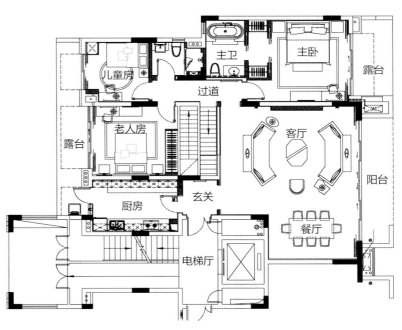

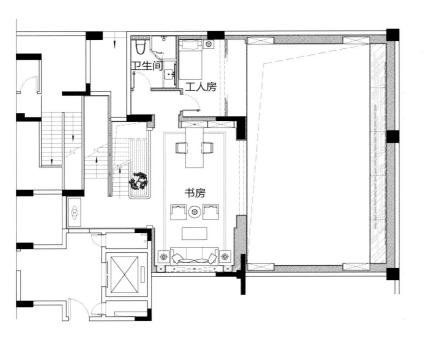

Formula | 方程式 ②

Flower arrangement is not only a kind of aesthetic art, but is beneficial to people's self-cultivation, and it is indispensable in interior design. In the white porcelain vase, there are several elegant white flowers, slanted or straight, they easily create an artistic feeling. Flowers that are put in the right place could provide a feast to one's eyes.

　　插花，不仅是一种美学艺术，可以使人修身养性，在室内设计中，更是不可或缺的一员。白瓷瓶中，几朵素雅的白花亭亭于枝条，或斜出，或直立，简简单单，便自成意境。恰当的插花，摆放在合适的位置，使人赏心悦目。

新中式风格 NEO-CHINESE STYLE

FENG XUEMEI
冯雪梅

项目名称：德清观云小镇

设计公司：枫桥定制

项目地点：浙江湖州

项目面积：110m²

主要材料：木质、竹、瓷器、布艺等

A Zen Style Retreat
世外禅居

DESIGN CONCEPT ｜ 设计理念

"…but after a few more score steps it widens into an open track. A wide plain was brought into view, dotted with houses in distinct order and full of good fields, beautiful ponds, mulberry, and bamboo. It was checkered with highways and paths between the fields…" This is the description of the paradise in Tao Yuanming's "Peach Blossom Spring". Everyone has a paradise in his heart. However, there is a real version of the "Peach Blossom Spring" in Mogan Mountain where it can realize all your dreams about paradise. In this landscape, only to put a villa of Chinese Jiangnan (regions in the south of the Yangtze River) style in the gentle slope around mountains can restore a yearning for the Jiangnan courtyard and the memory of the neighborhood life in the deepest heart of the Chinese. The case creates a "Neo-Chinese Style" indoor environment from the perspective of life, so that the family members can enjoy not only the natural environment, but also the life aesthetics.

"……复行数十步，豁然开朗，土地平旷，屋舍俨然，有良田美池桑竹之属。阡陌交通……"这是陶渊明《桃花源记》里关于世外桃源的描述，每个人心里都有一个世外桃源，然而，莫干山这处真实版的"桃花源"，可以满足你对世外桃源的所有幻想。

在这片山水中，在群山环绕的缓坡果岭之间，只有放下一栋栋江南中式院墅，才能还原中国人内心最深处对江南院落、对记忆深处的邻里生活的向往。本案从生活的角度去营造"新中式"的室内居家环境，让家人在观赏自然环境的同时，还能够享受生活的美学所在。

Formula｜方程式 ①

This case uses the natural materials to create an original and simple feeling and absorb the precipitation of years which is not impetuous, contrived and vulgar. The living room chose a large number of wood elements as the basic tone, uses the collocation of dark and bright colors to show the cultural heritage of homeowner, and to present a trace of quietness and tranquility. There is a corner of the natural space which can arouse the resonance of life. "I cannot live without bamboo" is a kind of lofty and elegant pursuit.

以自然材质，创造返璞归真的空间灵性。吸取岁月的沉淀，不浮躁，不做作，不庸俗。客厅里选择了大量木质元素作为铺底，深浅搭配既彰显了主人家的文化底蕴，又略带一丝安静恬淡。有一隅自然空间，唤起生活的共鸣。"不可居无竹"是一种清高雅洁的志趣。

Formula | 方程式 ②

Opening a window, the scene in front of eyes makes one's spirit feel comfortable. "With lyre-playing, chess, calligraphy, painting, tea and floriculture, the house is clean, quiet and full of Zen flavor." Just with a pot of tea and a book in this open Zen space, one can enjoy a quiet afternoon, return to the heart and forget the hubbub. Play Chinese zither to express the spirit of Zen, and seek friends who can really appreciate and understand your rhythm.

打开一扇窗，将眼前之景物，引申到精神层面的慰藉。"琴棋书画茶香花艺，朝夕净空满室皆禅。"一个人，一壶香茗，一本书，一个只属于自己的宁静下午的开放禅意空间，回归内心，忘却尘嚣。挥洒禅韵抚琴来，高山流水觅知音。

新中式风格 NEO-CHINESE STYLE

SHI SHAOFEN
施少芬

项目名称：广州从化方圆泉溪月岛B户型别墅样板房

设计公司：深圳市艺居软装设计有限公司

项目地点：广东广州

项目面积：250m²

主要材料：橡木、绒布、布艺等

摄 影 师：李健超

The Heart of Grass and Wood, the Home to Zen Flavor
草木为心 禅意之家

DESIGN CONCEPT | 设计理念

The overall design adheres to the traditional aesthetics which worships the original form to show the true colors of the original materials, presenting the unique texture of the material with precise grinding. This kind of filtering space effect has a calm, smooth visual surface, and it affects people's feelings, so that the city dwellers' potential nostalgia can return to the natural emotions and gain response. Space can merge with nature and borrow the external natural scenery to bring infinite vitality to the interior. The wood color furniture exudes a natural flavor, refreshing the soul like a breeze or a ray of sunshine.

整体设计上秉承传统美学中对原始形态的推崇，彰显出原始素材的本来面目，加以精密的打磨，表现出素材的独特肌理。这种过滤的空间效果具有冷静的、光滑的视觉表层性，却牵动人们的情思，使城市中人潜在的怀旧、怀乡、回归自然的情绪得到补偿。空间能与大自然融为一体，借用外在自然景色，为室内带来无限生机。草木色的家具散发出自然气息，目之所及的元素，如一阵微风，一缕暖阳，涤荡心灵，舒展身心。

Formula | 方程式 ①

Natural materials are widely used in decorations, with emphasis on elegant temperance and deep Zen flavor, rather than extravagant luxury and resplendent decoration, paying attention to the comfort and practical function of the space.

设计师将自然界的材质大量运用于装修装饰中，不推崇豪华奢侈、金碧辉煌的装饰，而以淡雅节制、深邃禅意为境界，重视空间的舒适性和实际功能。

Formula | 方程式 ②

Special emphasis is put on the natural texture and vitality in the selection of the indoor materials, which are mainly "grass" and "wood" so as to allow the space to have cordial exchanges with nature, achieving harmony between indoor and outdoor and creating an enjoyable residence.

室内选用材料上特别注重自然质感与生机，以"草、木"为主，让空间得以与大自然亲切交流，达到室内外和谐统一，居住其乐融融的境地。

新中式风格 NEO-CHINESE STYLE

WANG KUN
王坤

项目名称：大华·南湖公园世家

设计公司：王坤设计事务所

项目地点：湖北武汉

项目面积：180m²

主要材料：石材、黑色不锈钢、实木地板、木饰面、壁纸等

摄 影 师：王虎

Leisurely Landscape
悠然山水间

DESIGN CONCEPT | 设计理念

Chinese style is very particular about the level sense of space. In this case, Chinese-style screen, window lattice or wooden door are used in the places where the line of sight need to be cut off, showing the beauty of level of this Chinese style unit residence. With some simple modeling as the foundation, some Chinese elements are added, making the overall space more rich, large but not empty, thick and not heavy, dignified yet not suppressed.

The living room is the collision of traditional and modern living style, in which the designer employs modern decoration technique and furniture, combined with decorative elements of classical Chinese style to present a space atmosphere which is both ancient and modern. The elegance and quaintness of the Chinese style echo with the simpleness and grace of the modern style, so that the practicality of life and the pursuit of traditional culture are satisfied at the same time.

中式风格非常讲究空间的层次感，在需要隔绝视线的地方，使用中式的屏风、窗棂或中式木门工艺隔断，单元式住宅就展现出中式家居的层次之美。再以一些简约的造型为基础，添加中式元素，使整体空间感觉更加丰富，大而不空、厚而不重，有格调又不显压抑。

客厅是传统与现代居室风格的碰撞，设计师以现代的装饰手法和家具，结合古典中式的装饰元素来呈现亦古亦今的空间氛围。中式风格的古色古香与现代风格的简单素雅自然衔接，使生活的实用性和对传统文化的追求同时得到了满足。

In the use of colors, the bedroom adheres to the elegance and luxury of the traditional classical style, but the difference is that a lot of modern elements are added, showing the characteristics of fashion. In the choice of accessories, luxurious decorations are abandoned, expressing more fluently the essence of traditional culture. In order to add some warmth to the room, exquisite lighting and elegant paintings are used, making the entire room exude a modern atmosphere in the thick ancient rhyme.

居室在色彩方面秉承了传统古典风格的典雅和华贵，但与之不同的是加入了很多现代元素，呈现着时尚的特征。在配饰的选择方面更为简洁，少了许多奢华的装饰，更加流畅地表达出传统文化中的精髓。为了给居室增添几分暖意，饰以精巧的灯具和雅致的挂画，使整个居室在浓浓古韵中渗透出几分现代气息。

Formula | 方程式 ①

Landscape painting with a distant mood plays an important role in the shaping of the space temperament. In the living room, the painting sketches out a scene "vast river and boundless frost" with a few inks. Looking into the landscape, one is hit by constant and distant mood, just as the ancient people felt when they leaned on the railing. Such a simple Chinese landscape painting creates an appealing culture space.

意境悠远的山水画对整个空间气质的塑造有着重要的作用。寥寥几笔为客厅勾勒出一幅"寥廓江天万里霜"的景象，遥望山水，仿佛古人凭栏眺望，悠悠情思不断袭来。一幅简单的中国山水画，就营造了富有韵味的文化空间。

Formula | 方程式 ②

Lighting is one of the essential elements in the modern home furnishings. Simple and charming lighting decorations can add a lot of charm for the whole space, so the choice of lighting is also very important. In this case, the lighting embodies the traditional view of life: the way of square and round, reflecting traditional Chinese culture.

　　灯具,是现代家居中必不可少的元素之一。简洁富有韵味的灯具装饰居室,便可以为整个空间增添不少魅力,所以灯具的选择也至关重要。该案例中的灯具,均融入了传统的处世理念——方圆之道,可以让人从中感受传统文化。

新中式风格 NEO-CHINESE STYLE

ZHANG WEILUN
张炜伦

项目名称：北京电建泷悦长安

设计公司：卡纳设计

项目地点：浙江杭州

项目面积：417m²

主要材料：木、金属、玻璃、布艺等

摄 影 师：Christian Chambenoit 常易

Oriental Impression
印象东方

DESIGN CONCEPT | 设计理念

The magnificent temperament of the Tang and Song Dynasties deserves a modern interpretation. The Orientals like tea and pursue extraordinariness. Full of passion for modern life, they have their own clear judgments on the understanding of things and they pursue the quality of life. Through the careful operation of the material and the proportion of cutting, this case hopes to create an impression of the contemporary Oriental residence, which can carry the owner's calmness and self-confidence towards life and convey the infinite leisure and tranquility.

The space is calm and low-key, and the combination of flowers and trees adds a touch of interest to it. In the line of sight, the classic echoes with the modern while the warm and moist materials restore the most pleasant feelings, exuding a rich space charm in the light of the sunshine.

唐与宋的盛世气度，理所应当的现代演绎。东方情节，喜欢茶道，追求非凡，对当下现代生活充满热爱，对事物的认识有着自己清晰的判断，追求生活品质。本案通过精心的材料运营和比例切割，希望能创造出一个印象东方的当代住宅，它可以承载屋主对生活的从容与自信，并且能传达出无限的闲适与宁静。

沉稳低调彰显气质，浪漫的花与树组合，平添几许情趣。视线所及，古典与现代相互映衬。触手所及，温润材质还原最惬意的情怀，在光的映照下焕发出丰盈的空间意趣。

Formula | 方程式 ①

Blanking is a very important artistic expression in home design. In this project, designers also adhere to the use of Oriental traditional artistic ideas to deal with the "action and inaction" in the space, where the real is carefully created while the virtual is cleverly blank, making the real and the virtual complement each other.

留白是家居设计中一个极为重要的艺术表现手法。在本项目实践中,设计师也坚持运用东方传统的艺术思路来处理空间中的"有为和无为",凡实处则需要精心营造,凡虚处则巧妙地留白,虚实相生,相得益彰。

Formula | 方程式 ②

The use of soft decoration creates a sense of strength in the air. In the space whose basic tone is black and white, the colors such as dark green, gold, purple, blue are like the finishing touch, indispensable and full of vitality. At the same time, the space brings more beautiful things to a natural compatibility by the use of each other's relationship, showing the temperature of life.

利用软装，营造充满力量感的空间。在以黑白灰为基调的空间里，墨绿色、金色、紫色、蓝色有如画龙点睛一般，不可或缺，焕然生机。同时它将更多美好的东西，运用彼此的关系，自然相容，展现生活的温度。

新中式风格 NEO-CHINESE STYLE

PENG ZHENG
彭征

项目名称：佛山岭南天地·艺术收藏之家

设计公司：广州共生形态设计集团

参与设计：梁方其、陈泳夏

项目地点：广东佛山

项目面积：390m²

主要材料：浅色木饰面、白色烤漆板、天然麻质墙布、石材、水泥砖、黑色不锈钢、大理石、玫瑰金、玻璃、皮革等

Home of Art Collection
艺术收藏之家

DESIGN CONCEPT | 设计理念

Foshan has a rich historical and cultural heritage, and a large number of historical heritage buildings in the Zumiao Donghuali area have become the city's representatives of traditional culture essence and epitomes of this historical and ancient commercial city.

Foshan Lingnan Tiandi is located in the center of Donghuali area in Chancheng District, Foshan City. The project covers an area of 65 hectares, with a total construction area of 1.5 million square meters. Adjacent to Donghuali — the famous national unit for protection of cultural relics, and sitting on the lofty position of the city, it is a very rare traditional Chinese rich area with profound cultural heritage, and it preserves one of the most complete ancient building groups in China.

佛山拥有丰富的历史文化底蕴，其中祖庙东华里片区大批内涵丰富的历史文物建筑已成为这座城市传统文化的精粹代表，是佛山千年历史名城和古代商业重镇的缩影。

佛山岭南天地位于佛山市禅城区祖庙东华里中心地段。项目占地面积达65公顷，总建筑面积达150万平米。毗邻全国有名的国家级文保单位东华里，坐拥全城景仰的崇高地位，是中国极少数典藏深厚历史底蕴的传统富人区。岭南天地保存了中国其中一个最完整的古建筑物群。

Lingnan Tiandi·Jing Xuan is located in the North of Park Royale Phase III, and East of Donghuali Art Gallery. The project covers an area of about 17,800 square meters, with a total construction area of about 24,600 square meters. Design concept of this district is to arrange low-density residential buildings with green landscape, aiming to create a superior residential area. On the roof there are green plants watered with automatic irrigation, which is in accordance with Lingnan Tiandi's environmental protection concept.

岭南天地·璟轩位于三期御苑北面，东华里艺术长廊东面。项目占地面积约1.78万平方米，总建筑面积约2.46万平方米，小区主要以精心布置低密度住宅，并加以小区绿化，营造优越的住宅小区为设计概念。楼顶天面有绿化植物，浇灌采用自动浇灌，延续岭南天地一贯建筑环保概念。

The difficulty of the design of this case is that the original four-storey villa was split into two units for sales. Thanks to the home garden and the basement, the first and second floor are very marketable while the third and fourth floor are not so popular due to inconsistent life function. In view of the rich commercial formats and artistic atmosphere of the ancient district, the designers abandoned the standard scheme of conventional residential villa and defined the space to an art workshop closely connected to art collection, giving the space integrated functions of work, exhibition, residence and social contact.

　　本案设计的难点在于原本为一栋四层的别墅被拆分为上下两个单元销售，一二层由于带有入户花园及部分地下室，使得销售非常理想，相比之下，三四层则显得生活功能不够完整连贯。设计师另避蹊径，没有遵循常规住宅别墅的标准方案，而是结合古城区当地浓郁的商业业态和艺术氛围，将空间定位于与收藏艺术息息相关的艺术工作坊，并赋予空间工作、展览、居住、社交等复合功能与故事性。

Located on the second floor, the reception area is converted into a seven-meter-high vestibule by adopting a bold hollow handling, which not only makes the entrance space magnificent and grand, but introduces the natural light on the fourth floor into the second floor. The two-layer high decorative partition shelf not only serves as an entrance of the hall, but cleverly connects the upper with the lower space. Materials such as stone lions, long tea table, green moss, white sand stones create a quiet and distant artistic atmosphere together with the light.

位于二层的会客接待区通过大胆的中空处理，改建而成的七米挑高前庭，不仅使入户空间恢弘大气，亦将四层的自然光引入二层，两层高的装饰隔断柜不仅成为入口的玄关，亦将上下空间巧妙连接，石狮子、长茶台、绿苔藓、白沙石……材质与灯光营造出宁静、悠远的艺术氛围。

The artist's studio is on the third floor in which the open space is divided into a ceramic workshop and an appreciation area through decorative partition shelf with a lounge in addition. This is a small gathering space for various art salons and parent-child activities where every book, a bunch of flowers and a piece of artwork not only embody a historical and interesting adventure, but also an imprint of time and humanity. Here, we can return to the most true status of life and keep the original heart, and appreciate the aesthetics of life while listening to the rain and watching the flowers.

进入三层的艺术家工作室，开放式的空间通过装饰隔断柜将空间区隔成陶艺坊和鉴赏区，此外还配有一间休息室。这里还将是各种艺术沙龙和亲子活动的小型聚会空间，在这里，每一本书、一束花、一件艺术品，不仅意味着一次历史和趣味的探索之旅，也是时代与人文的印记，重要的是，我们回归最本真的生活状态，拥有最开始的那份初心，听雨落，看花开，悟出浅浅的生活美学之道。

There is a city, a history, a courtyard and a story here. The designer abandoned the deep and ancient Chinese style and re-interpreted the traditional evolutionary design concept with the current modern urban lifestyle. From the living room to the collection room, the workshop to the roof garden, a modern artistic space atmosphere close to the clients permeates everywhere, and a comfortable regional culture space is created.

一座城，一段历史，一个院子，一个故事。设计师摒弃深沉古韵的中式风格，用当下现代都市的生活方式重新诠释传统进化的设计理念，从会客厅到收藏室，从工作坊到屋顶花园，无处不营造出让客户亲近现代艺术的空间氛围，创作了一个舒适的区域化人文空间。

Formula｜方程式 ①

Light wood finishes match with natural linen and gray stone, creating a warm and peaceful background space. From the collection display shelf to the backlit screen, every detail echoes with the art furnishings and complements each other.

浅色木饰面、天然麻料与灰色石材的搭配，创造出温润平和的背景空间，从收藏展架到背光屏风，每一处细节都与艺术陈设互为映衬，相得益彰。

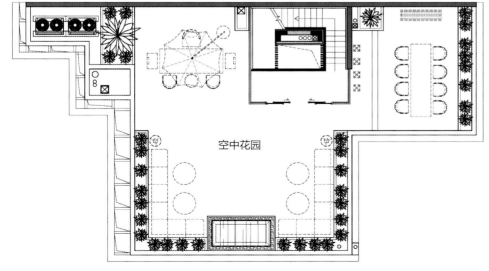

空中花园

Formula | 方程式 ②

On the upper floor, the grass house introduces warm sunlight into the room. The garden overlooks the skyline of the ancient city and huo'er walls (walls of an ancient building in Lingnan area) with green bricks and gray tiles. Throwing a private party here, one can not only feel a sense of history, but a sensation of nostalgia and comfort.

顶层用玻璃屋将温暖的阳光引入室内，空中花园可以眺望古城的天际和青砖灰瓦的镬耳墙，在这里举行一场私人聚会，不仅有几分历史的穿越感，更多了一丝怀旧与惬意。

NEO-CLASSICAL STYLE
新古典风格

CHEN BING
陈兵

项目名称：横店别墅

项目地点：浙江横店

主要材料：大理石、墙纸、布艺、水晶、地毯等

Low-key Interpretation of Luxury Temperament
低调演绎奢华气质

DESIGN CONCEPT | 设计理念

"The principle of simplicity" is a concept worshiped by modern design which extracts emotional inspiration from life and uses rational design techniques to break the inherent model of each space and create the most comfortable, natural and relaxed form through the design, so that a perfect home is achieved.

In the overall style of this case, unnecessary decoration is abandoned while clear lines, simple and neat models are adopted so as to integrate the classical luxury into the space, presenting the most natural state of life in a low-key way. The collocation of black and gray restores the most exquisite classical fashion through the aesthetic of smooth lines. The warm beige of marble constitutes the basic tone of the whole space together with the dark coffee, which complements with dark, magnificent and warm. The sparkling crystal chandeliers are another kind of romantic aesthetic.

"去繁从简"是现代设计需要崇尚的理念，从生活中提取感性的灵感，运用理性的设计手法，让每一个空间打破固有的模式，通过设计达到最舒适最自然最放松的形态，这便是一个完美的家。

本案设计在整体风格上，摒弃多余的装饰，化繁为简，线条明朗、造型简洁利落，将新古典的奢华转化为内在，以低调演绎的方式，呈现最自然的生活状态。黑色与银灰色的搭配，通过流畅且具有美感的线条，还原最精致的古典时尚。暖暖的米色大理石与深咖色构成整体空间的基调，与黑色互补，大气温馨。闪耀的水晶吊灯，又是另一种浪漫唯美。

Formula | 方程式 ①

Luxury and elegance are blended together to give the owner a different visual enjoyment. Lush flowers are planted in the room, not deliberately, comfortable and smart, adding a touch of natural freshness to the space. To make the relationship between space and people more comfortable and harmonious, romantic feelings are injected in this space in which people can truly relax.

将奢华与优雅融合在一起，给屋主不同的视觉享受。将葱郁的花植引入室内，不刻意、舒适而灵动，又增添一份大自然的清新。让空间和人的关系更加舒适、和谐，在这个理性的空间注入浪漫情怀，得到真正的放松。

Formula | 方程式 ②

The bedroom space reflects a noble and cozy atmosphere with an appropriate layout. Designers make a reasonable division of different functional areas to avoid stacked complicated items and break slightly thin lines of the wood with the unique sensory effect of colors. The contrast of static and dynamic creates a noble and elegant comfortable feeling.

卧室空间体现出高贵、惬意的氛围，在空间的布置上做到张弛有度。设计师将不同的功能区合理划分，尽量避免过于繁杂的物品堆放，用色彩特有的感官效果打破木材略显单薄的线条。一动一静既产生对比，又营造出十足的高贵典雅的自在舒适感。

新古典风格 NEO-CLASSICAL STYLE

ZHANG MIAO
章苗

项目名称：置信原墅

项目地点：浙江温州

项目面积：280m²

主要材料：大理石、护墙板、水晶灯、油画等

Extreme Luxury, the Taste of Home
极致奢雅 演绎家的味道

DESIGN CONCEPT ｜ 设计理念

Home is not only a relaxing port, a safe stop and a melting pot of love, but also an important place for communication where the fun of a family lies in carefree conversations. Therefore, it's very important to create a comfortable and charming atmosphere with eye-catching colors and smart furniture.

This case is a graceful and elegant French residential space, and the designer's pursuit of life quality is a promise to the householder. Stylish and comfortable sofas, unique tea table, ingenious color collocation, all these reveal the designer's careful handling of details. Lots of fabrics on which the exquisite flower patterns are woven with elegant and simple layered texture exude extreme elegance and romance. The whole elegant beige texture presents a warm and fashionable interpretation of the magnificent space.

　　家是安全的港湾，能舒适地停靠，是爱情的熔炉，同时也是沟通的重要场所，畅快的交谈尽在住宅之乐中。如何用夺目的色彩与智能的家具装饰出舒适迷人的氛围，非常重要。

　　本案是一套富有情调的优雅法式住宅空间，对生活品质的追求是设计师给予户主的期许。时尚舒适的沙发，独具匠心的茶几，别出心裁的颜色搭配，每一个细节的把控都十分重要。大量的花卉布艺、灵秀的图案交接着素雅简洁的层次纹理，最佳搭配展露着极致的优雅与浪漫。整体米色系的淡雅质感，使美轮美奂的空间在温馨中进行着时尚的演绎。

Formula | 方程式 ①

Bold colors are used to create a quality living environment just as you want. In the scorching summer, a wonderful home color would surely make you feel cool. The beautiful and attractive dark green makes the whole space exude a fresh breath. The dark green wainscoting and wall covering are extremely cool and fresh. The dark green from the dense forest is the color closest to nature in home furnishing. No matter how hot it is outside, it is always cool and fresh inside, presenting a quiet and mysterious scene.

大胆运用色彩打造你想要的精品居住环境。如火的夏季，美妙的家居色彩是否让你感受凉爽？那美丽且引人注目的墨绿色，让整个空间都散发着新鲜的气息。墨绿色的护墙板，墨绿色的墙布，清凉无敌。墨绿色来自茂密森林，是家居中最亲近自然的色彩。不论外界如何炎热，这里总是一片清凉，总是幽静神秘。

Formula | 方程式 ②

The handling of details decides the quality of home furnishing. This case is in classical French style with exquisite art decoration, carefully selected fine material and warm lighting which exude an elegant and charming temperament. The designer hopes to create an elegant and witty home space so as to let it feel more like a haven. From that moment you step into the house, you can enjoy a breath of fresh air and appreciate the beauty and romance of the space.

细节的把控，决定家居品质。经典的法式造型，考究的艺术品摆件，精细的材质选择，每一样都是用心挑选，再结合灯光氛围的营造，散发着优雅迷人的奢华气质。设计师想要创造一个典雅风趣的家居空间，让家感觉更像一个避风港，从走入家门的那一刻，便可以尽情地呼吸，尽情地感受美丽与浪漫。

新古典风格 NEO-CLASSICAL STYLE

WU HEJIAN
吴和建

项目名称：大华西溪橙宫

设计公司：尚层别墅装饰

项目地点：浙江杭州

项目面积：800m²

主要材料：挂画、布艺、地毯等

Fairyland in the World, Flower and Bird Paradise
人间仙境 花鸟天堂

DESIGN CONCEPT | 设计理念

This case has an American neo-classical style. "All our projects, even when executed in a modern style, have a definite twist, so they don't read as cliche," says George Yabu. Based on this, this case breaks through the traditional American style with the integration of the most popular design elements and a bold fusion of stone, color and classical elements to bring forth the new through the old. At the same time, it follows the people-oriented design concept, implements every detail after repeated scrutiny and brings different materials together through the combination of lines and planes, so as to meet the client's demand. The use of new materials agrees with the overall layout.

本案为美式新古典风格，"我们所有的设计，即使现代风格的也有一定的曲折变化，所以，人们不会认为是陈词滥调。"乔治·雅布先生说。基于此，突破传统装美式风格的理念，融合了时下最流行的设计元素，大胆将石材、色彩与古典元素融合，推陈出新，同时遵循以人为本的设计理念，在满足客户需求的基础上，不断反复推敲，落实到每一处细节，考究完善每一处节点和收口，不同材质通过线与面的组合方式拼贴，新材料的运用与整体布局浑然一体。

221

The overall color of the interior space is mainly yellow, reflecting a light luxurious temperament. The white walls are decorated with marble paint. Unlike the conventional layout, the overall layout makes the whole space magnificent and ventilative with a revolving stair connecting the upper and lower area space reasonably, enhancing the elegance of the whole space. The overall tone of the space is consistent, and the perfect combination of the wall and the ground makes the light and luxurious atmosphere penetrate the whole space, reflecting that the harmony of humanism is followed in shaping the comfortable and elegant sense of the space. The space is rich in level change with materials of different shades of color, making the whole space very harmonious.

　　室内空间的整体颜色以黄色为主，体现轻奢风。墙面采用白色护墙，加以大理石漆装饰。整体布局与常规布局相区别，使得整个空间通透大气，旋转楼梯连通上下区域空间，合理又不失整屋优雅的主题。空间色调整体一致，墙面与地面的完美结合使得轻奢气息贯穿到整个空间，在塑造舒适雅致的空间感官的同时遵循了人本主义的和谐。空间中富有层次变化，材质不同色度的渐变，使整个空间非常和谐。

Formula | 方程式 ②

The integration of modern elements with classical elements is a problem that many designers are thinking about and deliberating on. Modern materials are used to present the classical temperament, yet how to choose the appropriate materials and decoration is a test to the designer.

现代元素与古典元素的融合问题是很多设计师都在思考、实验、推敲的问题。运用现代的材料，展现古典的气质，如何挑选合适的材料和装饰就考验设计师的眼光了。

Formula | 方程式 ①

Modern people yearn for nature, aspire to the art of life and pursue comfort and leisure. Therefore, how to integrate the image of "flower" into the living room naturally and cleverly is critically important. From the flower arrangement on the table, patterns on the furniture, colorful images on the carpet, to the flower and bird drawings on the wallpaper, all show the different shapes of the flowers. These floral images, not only enrich the color of the room space, but make the living environment lively.

现代人向往自然，追求着生活的艺术，追求舒适与惬意，所以"花"的形象如何巧妙自然地融入居室中显得尤为重要。桌上摆放的插花、家具上的花纹、地毯上五彩缤纷的花的图案、壁纸上的花鸟图等等，无一不在展示着形态各异的花卉。这些花卉形象，不仅丰富了居室的空间色彩，更活泼了屋主的生活环境。

新古典风格 NEO-CLASSICAL STYLE

PATRICK FONG
方振华

项目名称：绍兴嘉华馥园样板房

设计公司：方振华创意设计（杭州）有限公司

参与设计：郑蒙丽、王尚坤

项目地点：浙江绍兴

项目面积：740m²

主要材料：透空雕花隔断、银箔、博古架、抽象油画、窗帘、羊毛地毯等

Great Harmony
大同

DESIGN CONCEPT ｜ 设计理念

In this villa, a decorative staircase is used to link the basement to the second floor while hollow carved partition is employed to connect the first floor with the second floor, making the overall style luxurious and majestic. Based on the hard decoration, styles of different layers have been taken into consideration in the design of soft decoration: the first floor is luxurious and magnificent, the second floor has different characteristics, the basement is for both leisure and entertainment. As a result, the layers of different styles are connected with each other, exuding a low-key luxury. As the center of the whole villa, the first floor is majestic and luxurious in which guests will be impressed by the luxury of the hall and the magnificence and gorgeousness of the lobby when entering. Therefore, the furniture accessories used on the first floor are mainly crystal, silver foil, varnish, glass and stainless steel, the space still presents a gorgeous temperament even it has no extravagant embellishment.

别墅用装饰性楼梯作为地下一层到二层的联系，而一层到二层则用镂空雕花隔断联系，整体的风格奢华大气。在硬装这个大前提下，软装在设计时用分层的方式来设定方向，一层的豪华大气，二层的别具一格，地下一层的休闲娱乐，不同的区域有不同的风格设定，各有特色又相互联系且透露出低调的奢华。一层主要以大气、奢华为主，是整个别墅的中心，客人进门的第一印象就是门厅的奢华和大厅的大气华丽。因此一层所用的家具饰品以水晶、银箔、亮光漆、玻璃、不锈钢等来做装饰，没有过多的雕刻，却也不失华丽。

As a recreation area, the basement is built as a yacht club, serving the function of both leisure and entertainment.

There are three main rooms on the second floor. The master bedroom is in the taste of the hostess, elegant, gorgeous yet magnificent. The boy's room is dominated by black and white, highlighting the boy's masculinity in adulthood while the main tone of the girl's room is pink and purple, presenting the girl's innocence and naivety. Though with their own unique characteristics, these three rooms are gorgeous and generous in common.

地下一层是一个休闲区域，为它打造一个游艇俱乐部的氛围，既休闲又有娱乐性。

二层为三个主要房间，主卧是以女主人为主，优雅华丽不失大气；男孩房则以黑白为主，突出成年后男孩的阳刚之气；而女孩房以粉紫色调为主，凸显小女孩天真烂漫的特质，三个房间各具特色，总体都是以华丽、大气为主。

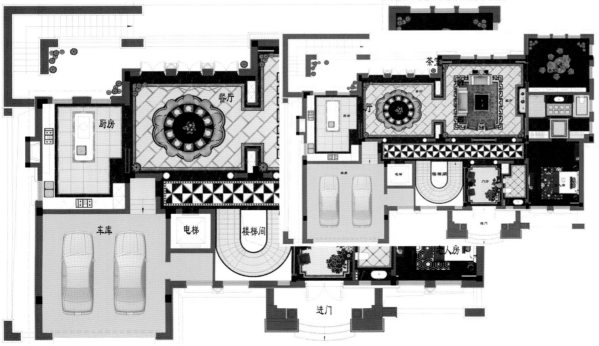

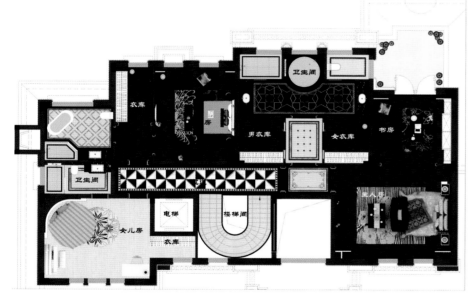

Formula | 方程式 ①

Wall decoration is a major key to interior decoration. Suitable decoration can not only beautify the space, but also add luster to the space. In this case, exquisite patterns and cravings on the living room wall exude an elegant temperament. Various textures on the bedroom walls express different life interests: the master bedroom is elegant and magnificent, the elder's room exquisite and stately, the girl's room graceful and concise and the boy's room straight and clear.

　　墙壁的装饰是室内装修中的一大要点，合适的装饰不仅能够美化空间，还能够为空间增光添彩。客厅墙壁上的精美花纹和雕饰，彰显优雅的格调。卧室墙壁形态各异的纹理，表达不同的生活韵味：主卧条理分明、淡雅大气，老人房花纹精细、沉稳大方，女孩房纹理优美、简洁清晰，男孩房直线平行、色彩分明。

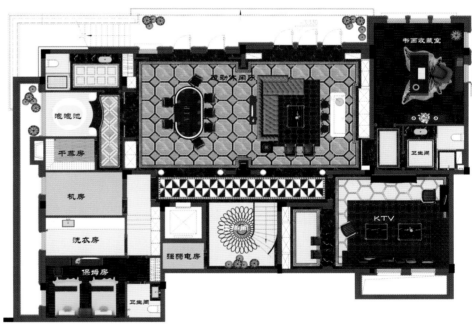

Formula | 方程式 ②

The appropriate interior decorations play a role in beautifying and highlighting the space. For example, the brightly-colored flowers enrich the color of the living room, ornaments of different shapes create a unique cultural atmosphere of the space and the bonsai placed in the open corner makes the living environment more natural and harmonious.

合适的室内装饰品能够起到美化和衬托空间的作用。颜色鲜艳的花卉，丰富居室的色彩，形态各异的摆件，营造空间独特的文化氛围。摆放在空旷角落的盆栽，使居住环境更加自然和谐。

AMERICAN STYLE 美式风格

美式风格 AMERICAN STYLE

FENGQIAO CUSTOM DESIGN
枫桥定制

项目名称：镇江东方诚园美式别墅

项目地点：江苏镇江

主要材料：大理石、布艺、壁纸等

Beating American Feelings
跳动的美式情怀

DESIGN CONCEPT | 设计理念

The American style shows a diverse and colorful international tendency, and integrates American people's human elements, such as freedom, liveliness and creativity. At the same time, American style pursues the quality of life and the return of sentiments, so that people can feel a "home" atmosphere everywhere. Based on its humanity, history and life, the American style has some unique characteristics, such as being implicit, simple, natural and concealing, which are not fully possessed by other nationalities in Europe, and it is these characteristics that make it still vitalized till today.

美式风格呈现出多元的丰富多彩的国际化倾向，融合了美国人自由、活泼、善于创新等一些人文元素。同时，美式风格追求生活品质和回归情怀，使人们每到一处皆能感受到"宾至如归"的气氛。以其独有的内敛、质朴、自然、深沉、人性、历史、生活的特质是欧洲其他民族所不完全具备的，也正是基于这些特质，使得美式一直到今天仍然充满着旺盛的生命力。

As a hospitality area, the overall feeling of the living room is bright yet not short of ancient flavor. The freedom and casualness of the American style can be felt from the sofa and its accessories. And there is a ubiquitous expression of the owner's pursuit of an exquisite life. As the heart of home life, the beautiful and generous dining room also attaches importance to practicability and functionality, in which the color and shape of the round dining table accord with the magnificence and calmness of the classical style.

客厅作为待客区域，整体感觉明快而不失古韵，从沙发和饰品的摆放上可以感觉到美式风格的自由和随意。业主追求精致的生活意愿无处不在地表达出来。餐厅是居家生活的心脏，不仅美观大方，而且在实用性和功能性上也同样重要，圆形餐桌的色泽和造型都遵循着古典风格的大气和沉稳。

Formula | 方程式 ①

The choice of jumping colors makes a perfect presentation of youth and casualness. Lemon yellow with gray blue is an eye-catching color match with one as bright as the sun while the other as cool as the sea. A stylish and elegant space gesture is created in the collision of colors of strong personality.

跳跃的色系选择，将青春与随性完美呈现。柠檬黄搭灰蓝色，是大自然教给我们的吸睛战术。一个灿烂如骄阳，一个清冷似海水。超强的个性碰撞中，打造时尚优雅的空间姿态。

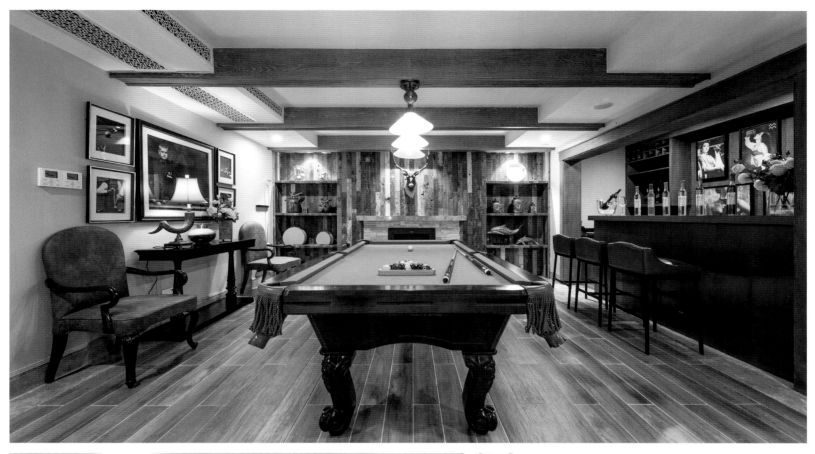

Formula | 方程式 ②

This case is particular about the openness of the spatial pattern and the creation of a warm atmosphere. Bulky furniture, comfortable though it is, is easy to cause a sense of space oppression, so in order to avoid this, the planning of spatial pattern is particularly important. The designer makes a clever use of the open layout, which not only creates a continuous and smooth atmosphere, but effectively increases the interaction among family members.

注重空间格局的开放性，温馨氛围的营造。大体量的家具沙发，虽然舒适性很强，但是也很容易造成空间的压迫感。为了避免压抑感，空间格局的规划就显得特别重要。设计师巧妙利用开放式，连而不断，既通透大气，又可以有效地增加家人之间的互动。

美式风格 AMERICAN STYLE

HE WENZHE
何文哲

项目名称：济南雪野湖别墅

设计公司：西盛建筑设计（上海）事务所

参与设计：唐瑶华，顾陈玮

项目地点：山东济南

主要材料：木质、布艺、铁艺、壁纸等

摄 影 师：朱沈锋

An American Mansion of Free Design
追求自在设计的美式美宅

DESIGN CONCEPT | 设计理念

Considering the client's life interests, the designers created a comfortable Minimalist house of American country style. The overall space is dominated by low-key white, gray and wood color to create good quality. The design emphasizes on "gray white", reserved and not showy, simple and comfortable. The use of the warm earth yellow in some parts makes the space exude warmth in gray.

设计结合业主生活情节，打造自在乡村美式极简住宅。整体以最为低调的白、灰、木三色为主，营造质感空间。设计重笔在"灰白"，含而不露，简约舒适。局部使用暖色系的大地黄，在灰盒子中渗透着温度。

In the overall layout, the design interprets the freedom and casualness of American style to the extreme. The proportion of division lines is appropriate, simple yet exquisite. The large window outlines the most beautiful scenery through which the bright light brings the warmth of sunshine inside. The designers aimed to create a high-end American mansion with the integration of leisure and pleasure.

在整体布局中，将美式设计的自由随性演绎得极为纯粹。空间线条的比例划分到位，简约又不失精致。偌大的窗，框出了最美的风景。明亮的光透过敞亮的窗户，将阳光的暖意带了进来。设计师意在打造一处集休闲与享乐为一体的高级美式豪宅。

Formula | 方程式 ①

Minimalism is an attitude and a kind of complex. In this case, the designers opposed stodgy design attitude, abandoned complicated design and simplified people's life to create a more suitable living environment. Under the open layout, a few simple tables and chairs with several fresh flowers can be the whole part of life.

极简，是一种态度，是一种情结。摒弃过多繁复的设计，反对刻板的形式设计态度，勿堆砌，给生活做减法，追求更适合人居住的生活环境。开阔的格局下，简约的几桌几椅，再摆上几束新鲜采摘的鲜花，便可以是生活的全部。

Formula | 方程式 ②

As the unchanged days make life boring and tasteless, a relaxed design is very important. A comfortable design is a creation of change, a collision between rigorous rational thinking and overflowing artistic inspirations. All the design in the concept of appropriate residence should be promoted, as it may insensibly influence our life habits, improve the quality of life and shape the taste of life.

学会放松设计。一成不变的生活造就了渐变无聊的日子,舒适的设计是一种改变的创造,是澎湃的艺术灵感与缜密的理性思维之间的碰撞。一切以适合人居为理念的设计都应该被推崇,它可以在不知不觉中影响着我们的生活习惯、提升生活品质,也塑造着生活品格。

LIAN JUNMAN
连君曼

项目名称：融侨观澜TLF-1

设计公司：云想衣裳室内工作室

项目地点：福建福州

项目面积：280m²

主要材料：仿古砖、文化石、实木、布艺、乳胶漆等

摄 影 师：周跃东

Notting Hill
诺丁山

DESIGN CONCEPT | 设计理念

Named after the film *Notting Hill* by the owner, this case has a warm American style with a touch of industrial flavor.

As the living room area is cramped, the designer extended the space to the balcony so as to put an additional individual sofa. This roof beam belongs to the original building, since a flat roof is too low, then two false beams are added for decoration. As a result, the cultural stone vault modifies the gap. The red carpet under the bay window is a shoes change cushion in the entrance. Considering the space layout, the kitchen was handled by an open plan while the living room is also used as the study with the balcony included into the room as a reading area. Balcony, changing room and bathroom are all included in the master bedroom, forming a large suite, gorgeous and magnificent. Out of the window is the river view, saving the problem of being peeped by people in the opposite buildings. The attic layout is arranged according to the skylight, where those spaces that are too low or dark have been planned for storage. The texture of fur decreases the cold industrial feeling. As the dining room is quite small, the designer extended the space to the elder's room. The wooden door and the balcony with culture stone give people a sense of affinity.

　　本案名称由业主亲自所起，来源于同名电影《诺丁山》。温馨的美式风格中，夹杂着些硬朗的工业味，是本案最显而易见的特色。

　　客厅面积局促，只能向阳台扩展，多摆下一个单人沙发。一根梁是原建筑的，封平屋顶太矮，于是多补两根假梁做装饰，文化石拱顶修饰空间里的落差。飘窗下的红色软垫是玄关换鞋凳。考虑空间问题，厨房采用开放式处理。起居室兼书房，阳台被包入室内作为阅读区。露台、更衣室和卫生间全部划入主卧内，形成大套间，看过去非常大气，且有气势。露台窗外是江景，视野独特，尽享美景。阁楼的布局围绕天窗安排，太矮或者光线不好的地方通通规划为储物空间。右边白色砖墙内是储物室，毛皮的质感冲淡工业味的冰冷。餐厅略小，便往老人房方向扩展了一些，木门和贴文化石的阳台，给人以亲和之感。

Formula ｜ 方程式 ①

The renovation of old house needs effective planning. The designer arranged the overall space layout according to the degree of importance. As the space is not wide enough, an open layout could give people a broad vision and a spacious feeling.

旧房改造中需要对空间进行有效的规划，按照主次原则，安排整体空间布局。若是空间不够宽敞，开放式的布局则能给人以开阔的视野和愉悦感。

Formula | 方程式 ②

The use of culture stone gives the room space a different flavor. It is widely used in vault, balcony and so on, creating a sense of history and a more sedate temperament. At the same time, it gives the host aesthetic enjoyment.

　　文化石的使用赋予了居室空间不同的风味，拱顶、阳台等大面积的文化石营造出历史感，更显沉稳气质，同时，也给予主人审美上的享受。

KANG YUAN, MENG MENG
康源、孟孟

项目名称：沈阳听雨观澜

设计公司：幸福格色空间设计

项目地点：辽宁沈阳

项目面积：330m²

主要材料：木质、铁艺、壁纸等

A Gentle Remembrance
一段温柔的忆想

DESIGN CONCEPT | 设计理念

Every space design is like a wonderful journey of life. In this case, the host and hostess are calm, kind and clever. During the entire design process, the designers felt a warm and clean stream gently flowing in their hearts.

This is a three-storey villa with a not too small basement. There are four members of three generations in the family. The villa is very spacious with many rooms, so the designers retained the overall spacious effect and adopted the fresh and natural American country style. At one corner by the window in the living room, there is a comfortable sofa on which the homeowners could take a nap under the sunshine, read a book, listen to the music or have a sweet dream...

每一个空间设计，都是一段美妙的人生之旅。本案的男女主人，性格沉静，善良聪明，整个设计过程，像一股暖暖的清流，在心底温柔地流淌。

这是一栋三层别墅，外加一个不算小的地下室。三代四口人，空间十分宽敞。房间很多，所以设计保留了整体宽敞的效果，风格定位为自然清新的美式乡村风。所以特地在客厅靠窗的角落，布置了一张舒适的休闲沙发，屋主可以在某个风吹帘动的午后，拥一怀暖阳，读一本书，听一段音乐，或者做上一个美梦……

Formula | 方程式 ①

Handmade furniture can not only increase retro features but guarantees its quality. The handmade color-processed solid wood cabinet echoes perfectly with the retro bookcase in the living room, highlighting the continuity of the space.

手工打造，既可以增加复古特色，又能保证家具品质。现场手工打造的套色实木橱柜，效果惊艳，与客厅空间的复古书柜完美呼应。而且同时可以为各空间的延续性增添亮点。

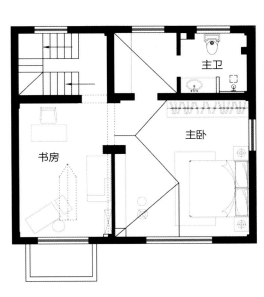

Formula | 方程式 ②

The designers made an ingenious combination of the homeowners' hobbies and space planning. For example, the master bedroom is a big suite linked with the study which serves as a place for storage, resting and reading, catering to the owners' demands of reading and working. The basement is set as a recreational fitness room, where there is a ping-pong table that not only provides entertainment, but increases the flavor of life.

巧妙将主人的爱好与空间规划结合。主卧是一个大大的套间，与之相连的是一个集收藏、小憩、阅读功能为一体的书房，可以很好地满足主人的阅读或者工作需求。地下室则设为休闲娱乐、健身空间，放置了一张全家都爱的乒乓球台，既娱乐又增加生活的气息。

美式风格 AMERICAN STYLE

GRACE KWAI
桂峥嵘

项目名称：中海紫御豪庭

设计公司：上海桂睿诗建筑设计咨询有限公司

项目地点：上海

主要材料：大理石、模板、布艺等

摄 影 师：ingallery™ 逆风笑

Fashionable Pulsations in Shanghai, Listening to Free Voices
时尚魔都脉动 聆听自由心声

DESIGN CONCEPT | 设计理念

As the first pure residential project in Changfeng ecological business district in Shanghai, the Amethyst tries to create scarce villas and big high quality flats. The Australian landscape designs create a classical European garden. The comfortable and rich skyline softens cultural connotations of the city.

The house belongs to Grace's friend, so she doesn't need to take such great pains in the project position like she did for her company's projects, but can be more free and casual. But if we say the design position of this project is random, it is obviously inaccurate. The design position of this home is closer to modern urban style which combines with spirits of modern Shanghai and continues European tone of the whole houses of the Amethyst. Grace matched quiet gray and warm milky white with exquisite technique of expression of multi-materials to deeply explore the life belief pursued by elites in modern city and their rich inner world.

　　中海紫御豪庭，这个魔都某一年的地王作为长风生态商务区首个纯住宅项目，着力打造稀缺性别墅和高品质大平层，来自澳洲的景观设计营造了一个欧洲古典风情园林，而舒缓丰富的天际线，又柔和着城市人文的底蕴。

　　作为朋友的家，设计师并没有像通常对待公司的项目那样在定位上殚精竭虑，更多的是随性和从容的挥洒。但是，如果说本案的设计定位就是信马由缰，那显然是不准确的，这个家的设计定位更贴近现代都市格调，将摩登魔都的精神贯穿其中，也延续着中海紫御豪庭楼盘整体的欧式格调。以沉静的灰调和宁馨的乳白，结合多元材质的细腻表现手法，深入探索当下现代大都市精英圈层执着的生活信念和丰富的内心世界。

When elaborately designing her friend's house, Grace transformed the classical European layout and details to a modern, neat and concise style. She skillfully combined Oriental and Western, classical and modern cultural essences to present a fashionable yet not vulgar, uninhibited and unrestrained multi-level three-dimensional individualized space which has the cross-sectional beauty after the opening of Shanghai to the world.

在精心打造朋友家的时候，设计师同样秉承从传承古典欧式的格局和细节到现代洗练爽洁的摩登都市范儿，游刃有余地糅合东方和西方、古典和现代的人文精髓，呈现出时髦却又不媚俗、不羁而又内敛的多层次立体个性空间，兼具上海开埠以后各个横断面的美。

The designer provided comfort, leisure and nature to the owners' mornings and nights, which could touch their souls.

设计师赋予空间这种安闲自在，这种天籁希声，伴随着她的晨昏，熏染着别墅主人的灵魂。

Formula | 方程式 ①

The designer chose calm and tranquil gray, warm and mild milky white to outline the rich and lively details under the concise and neat modern frame. The natural grains and colors of materials harmonize and correspond, or contrast and collide with each other. The designs start from exquisite observation and rich personal experience, combine the past golden times and modern urban intentions of Oriental and Western times and endow the project with fashionable metropolitan atmosphere by light luxurious and modern visual experience. The authentic American dining table from the 17th century foils a warm, amiable and elegant temperament of the dining environment, which makes the entire atmosphere more relaxing and leisure and brings heavy sense of history and cultural deposits. Pure color leather and cloth sofas, abstract art painting, Victorian porcelain display cabinet, phonograph, red wine and sheepskin pad, all these elements with strong feelings are used together, creating a modern and classical feeling and building an individualized space which is cross-boundary, unrestrained but can make people enjoy its connotations calmly.

选用沉稳宁静的灰调和温馨柔和的乳白，在简洁的现代造型框架内，勾勒出丰满灵动的细节表情。多种材质的天然纹理色泽，或和谐相应，或对比碰撞。设计以细腻的体察和丰富的个人经验作为出发点，复合了东西方往昔流金岁月与现代都市的意向，以轻奢摩登的视觉体验令全案浸润在时尚的大都会气息中。17世纪风的真品美式餐台烘托出就餐环境温暖亲和又典雅的气质，在使得空间整体气氛更为轻松自在的同时，带来历史文化积淀的厚重感。色彩纯一的皮革和布艺沙发、抽象艺术画、维多利亚风格的瓷器陈列橱柜、唱机、红酒、羊皮脚垫……这些带有浓厚情怀的元素调配在一起，摩登和古典交融的感觉油然而生，构建出了一个有着些许跨界的不羁却可以让人沉下心来细细品味其内涵的个性空间。

Formula | 方程式 ②

The master bedroom and guest room use neat and wide layout to present the owner's pursuit of freedom and unrestrained personality. The maximum use of space scale and unlimited collocation of interior day-lighting and artificial light prove this kind of truth that returning to the original nature is the most suitable belonging after prosperity and the final design position is to make people feel the most comfortable existence.

以洗练宽阔的布局展现主人追求自由、不受羁绊约束的个性。主卧和客卧空间尺度的最大利用，室内采光和人造光源的配合不受局限，无不证明着这样一个道理：返璞归真是历经繁华后最合适的归属，家最终的设计定位其实就是一个让人感到最惬意放松的所在。

SOUTHEAST ASIAN STYLE
东南亚风格

PENG JUAN
彭娟

项目名称：融创紫泉枫丹别墅样板间

设计公司：重庆于计设计

项目地点：重庆

项目面积：300m²

主要材料：大理石、金属、软包、挂画、藤条等

摄影师：刘星昊

Dazzling Rainbow
彩虹之炫

DESIGN CONCEPT | 设计理念

Southeast Asian style combines the features of Southeast Asian island and exquisite culture taste, with primitive nature, bright color and handicrafts as its main characteristics. The wooden symmetry structure is dominated in the house, creating a strong tropical flavor. In terms of color, the primary thick brown is interspersed with dark green and tinged with bright purple, which is warm and natural yet enthusiastic and gorgeous. Through the deconstruction of the peacock elements, the designer presented the leisure resort flavor of Jinyun Mountain incisively and vividly.

东南亚风格，它是一个东南亚民族岛屿特色与精致文化品位相结合的设计。原始自然、色泽鲜艳、崇尚手工是它的主要特点。在造型上，以对称的木结构为主，营造出浓郁的热带风情。在色彩上，以厚实的棕色为主，暗绿为辅，局部点缀艳丽的紫色，自然温馨中不失热情华丽。在本次融创紫泉枫丹联排别墅项目中，设计师们通过解构孔雀元素，将缙云山脚下的度假休闲风情打造得淋漓尽致。

Formula | 方程式 ①

The room is dotted with handmade decorations of different shapes, exuding an exotic flavor. The warm yellow light, decorations with unique flavor, gentle living atmosphere make people can not help but indulge in it.

  造型各异的手工装饰品，分布在居室中，让每一处的异域风情都自然流淌在居室空间之中。暖黄的灯光，风味独特的装饰，柔和的居室氛围，让人不禁想要沉醉其中。

Formula | 方程式 ②

The bright-colored fabrics enrich and soften every inch of the space, dazzle people's eyes, bring surprises, enjoyment and imagination with its exquisite patterns.

　　色泽鲜艳的布艺装饰，丰富、柔和着居室的每一处，不仅让人目不暇接，不断迎来惊喜，而且其花纹的精致独特也给人以美的享受和联想。

东南亚风格 SOUTHEAST ASIAN STYLE

CHONGQING YOKY DESIGN
重庆于计设计

项目名称：融创金开融府别墅样板间A户型

项目地点：重庆

主要材料：布艺、软包、挂画、藤条等

Home of Rainforest
热带雨林之家

DESIGN CONCEPT | 设计理念

In this case, Southeast Asian style with a strong exotic taste is adopted. The essence of the whole space lies in the use of colors and the collocation of soft decoration. In terms of visual effect, it gives people a feeling of shock and impact as if they are in the rain forest; in the aspect of space atmosphere, it creates an exotic atmosphere with the contrast of different cultures and the display of primitive ecology. The dominant hue of the space is warm while bold and bright colors are applied to soft decoration matched with the natural wood color, giving its residents a warm home feeling in a strong exotic atmosphere.

本案选用异域风情浓重的东南亚风格，整个空间的精髓游离于色彩的运用和软装的搭配之间，视觉上给人进入热带雨林的震撼击之感，空间氛围上追求民族文化的差异与原生态写意的感觉。整体为暖色调，软装的用色大胆明艳，与原木色搭配，在沉稳大方中透出活泼灵动，在浓郁的异域风情中感受到家的温暖。

Formula | 方程式 ①

In this case, the decorations are one of the elements that can make people feel exotic. Hand polished crafts of different characters seem to tell stories that have long been popular, which not only enrich the space but present the unique charm of the host.

　　装饰摆件是此案例中最能让人感受到异域风情的要素之一。手工打磨、造型各异的人物形象的工艺品，仿佛在诉说着一个个流传已久的故事，不仅丰富着空间，也在传达着主人独具特色的魅力。

Formula | 方程式 ②

Texture is the most striking element of this case, the linear texture on the floor, the geometric pattern on the ground, the pattern on the ceiling, the flower pattern on the curtains, the ink-splashing texture on the porcelain, all of them have characteristics of their own yet display their charms for the space together.

　　纹理，是此案例中特别引人注意的元素。地板上的流线型纹路、地毯上的几何造型、天花板上的图案、窗帘上的花纹、瓷器上泼墨般的纹理……各有特色，各成一景，却又共同为空间增添无限魅力。

东南亚风格 SOUTHEAST ASIAN STYLE

KENNETH KO
高文安

项目名称：海南华凯·南燕湾海崖别墅H户型

设计公司：深圳高文安设计有限公司

项目地点：海南万宁

项目面积：478m²

Simplicity out of Complexity, Deep Love for Nature
出繁入简 厚爱自然

DESIGN CONCEPT | 设计理念

Only after we have visited the world can we know that it's difficult to have a life that is close to cliff and sea. In the large city which is occupied by steels and cements, no matter how we talk about nature, we still can't fully enjoy it. But this cliff villa is rooted in the picturesque scenery of Ocean Bay, and it has natural views of the mountain and sea. The designer created a perfect sea resort elaborately and followed the nature of life.

The interior retains the natural grain pattern of the primordial trees in the rain forest, revealing the plain and fragrant vegetation of Southeast Asia, but also presenting the closest affection between oriental culture and nature. The steady wood color collides and blends with the dynamic sea blue. The still mountain and the dynamic sea create a decent elegance. Bedroom is close to the infinite swimming pool in the private garden. Beside the pool, the beautiful flowers are flourishing. In the distance, the green mountain and vast ocean provide an open and transparent view. The designer used wood droplights and shells to decorate the interior, creating a leisurely feeling between the green mountain and ocean and achieving a perfect fusion of the nature and daily life.

世界阅尽，方知人生难得一片悬崖藏海。在被钢铁水泥圈占的大城市，不管我们怎么奢谈自然，都是自然的门外汉，不得其道。而根植于南燕湾隽秀风光中的海崖别墅，开门见山，推窗观海，设计以匠心打造完美度假海居，呵护生活的道法自然。

室内保留了雨林原生树木的天然纹理，透出东南亚的朴实无华与草木芬芳，也带有东方文化中与自然相亲的情愫。沉稳的木色与灵动的海蓝色碰撞交融，山的静气，海的灵气，酝酿出不俗的风雅之气。卧室靠近私家花园的无边际游泳池，临水而居，池边鲜花争妍斗艳，远处山青海阔，视野开阔通透。室内以枯枝吊灯、贝壳装饰，点缀逍遥于山林碧海之间的悠然情怀，自然与生活起居完美融合。

Living in large cities where the land costs are extremely high, we hope that every inch of space is utilized adequately. But here, the design outlines the fresh and elegant space tone in a simple way with more "useless" blank space where people do not possess but wait the natural scenery and interesting details of their own life to fulfill it. Under the natural context, the modern Southeast Asian style is full of passion and vitality. The natural and peaceful environment contains people's pursuit of local spirit, and the changing colors in the quiet living space create not only an experience of rich visual perception but also a feeling of surprise and romance in the psychological perspective. Dwelling here is the beginning of a wonderful life.

在寸土寸金的大城市生活，我们希望每一寸空间都物尽其用。但在这里，设计只以简单的手法勾勒清新淡雅的空间基调，更多"无用"的留白，人不去占有，自然风光和生活本身的有趣细节会去填满。自然语境下的现代东南亚，并不缺激情与活力。自然平和中带有本土的精神追求，清静的生活空间中装入多变的色彩，带来的不仅是丰富的视觉感受，更在心理层面给人惊喜、浪漫。栖居于此，是多姿多彩生活的开始。

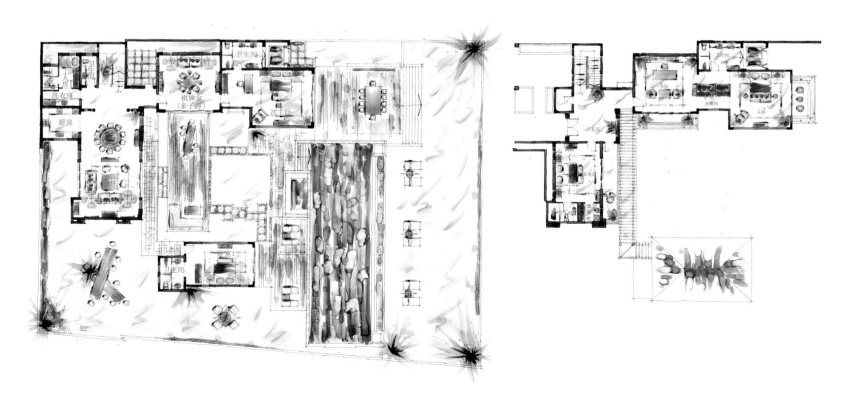

Slope roof in wood structure and linen veneers create a simple, comfortable and modern Southeast Asian style. International design elements, such as hollow wood carved screen, patterned Turkish pillows and carpet, are applied in it; together they present a combination of Chinese and Western cultures. The designer used top luxury private facilities to create an ultimate living space that is both "prosperous and quiet". Taking a bath in front of the window, the homeowner can enjoy the symphony of waves and mountain wind, and the beautiful sunset. Everything is beautiful and wonderful.

木结构的坡屋顶，麻布料的内饰面，印染出简单、舒适的现代东南亚风情。镂空的原木雕花屏风，土耳其做旧风格的拼接抱枕与地毯，国际特色的设计元素带来中西合一的文化格局。以奢极若简的顶级私属配套，打造"出得繁华，入得幽静"的极致生活。当你在海景窗前泡澡时，海浪与山风的交响，美得让人窒息的日落，一切美好都会扑面而来。

Formula | 方程式 ①

The natural texture penetrates into every detail of design. Tidal texture of the floor and wall tiles show the surging ocean atmosphere pouring into the room from the bay. Bright yellow decoration furnishings and beige linen furniture bring the warm texture of sunny beaches. The use of rattan, bamboo, wood carving and other Southeast Asian elements leads a simple but elegant resort life style.

自然的肌理渗透到设计的每个细节。潮汐纹理的地板与墙砖，澎湃的海洋情调从海湾涌进室内。明黄色装饰陈设，米白色麻布料家具，带来阳光沙滩的温暖质感。藤编、竹编、原木雕花等东南亚元素的运用，引领简约又不失精致的度假生活方式。

Formula | 方程式 ②

In the soft decoration, the bright colors which are inspired from the red and green rain forest and the handicrafts which imitate Southeast Asian ethnic costumes skills create an ocean resort with primitive characteristics and leisure atmosphere.

软装饰鲜艳的色彩，提取自花红草绿的热带雨林，工艺上参照了东南亚少数民族服饰的技巧，呈现出一个带有原始特色及休闲氛围的海洋度假场所。

图书在版编目（CIP）数据

设计方程式：样板空间风格赏析 / 深圳视界文化传播有限公司编． -- 北京 : 中国林业出版社，2017.9
ISBN 978-7-5038-9251-6

Ⅰ．①设… Ⅱ．①深… Ⅲ．①室内装饰设计 Ⅳ．① TU238.2

中国版本图书馆 CIP 数据核字（2017）第 207253 号

编委会成员名单
策划制作：深圳视界文化传播有限公司（www.dvip-sz.com）
总 策 划：万绍东
编　 辑：杨珍琼
装帧设计：黄爱莹
联系电话：0755-82834960

中国林业出版社 • 建筑分社
策　　划：纪　亮
责任编辑：纪　亮　王思源

出版：中国林业出版社
（100009 北京西城区德内大街刘海胡同 7 号）
http://lycb.forestry.gov.cn/
电话：（010）8314 3518
发行：中国林业出版社
印刷：深圳市雅仕达印务有限公司
版次：2017 年 9 月第 1 版
印次：2017 年 9 月第 1 次
开本：235mm×335mm，1/16
印张：20
字数：300 千字
定价：428.00 元（USD 86.00）